AI and Blockchain Optimization Techniques in Aerospace Engineering

U. Vignesh
Vellore Institute of Technology, Chennai, India

Rahul Ratnakumar
Manipal Institute of Technology, India & Manipal Academy of Higher Education, India

Abdulkareem Sh. Mahdi Al-Obaidi
Taylor's University, Selangor, Malaysia

A volume in the Advances in Mechatronics and Mechanical Engineering (AMME) Book Series

IGI Global
PUBLISHER of TIMELY KNOWLEDGE

Published in the United States of America by
IGI Global
Engineering Science Reference (an imprint of IGI Global)
701 E. Chocolate Avenue
Hershey PA, USA 17033
Tel: 717-533-8845
Fax: 717-533-8661
E-mail: cust@igi-global.com
Web site: http://www.igi-global.com

Library of Congress Cataloging-in-Publication Data

CIP Data in progress

Title: AI and Blockchain Optimization Techniques in Aerospace Engineering

ISBN: 9798369314913

This book is published in the IGI Global book series Advances in Mechatronics and Mechanical Engineering (AMME) (ISSN: 2328-8205; eISSN: 2328-823X)

British Cataloguing in Publication Data
A Cataloguing in Publication record for this book is available from the British Library.

For electronic access to this publication, please contact: eresources@igi-global.com.

Advances in Mechatronics and Mechanical Engineering (AMME) Book Series

ISSN:2328-8205
EISSN:2328-823X

Editor-in-Chief: J. Paulo Davim, University of Aveiro, Portugal

MISSION

With its aid in the creation of smartphones, cars, medical imaging devices, and manufacturing tools, the mechatronics engineering field is in high demand. Mechatronics aims to combine the principles of mechanical, computer, and electrical engineering together to bridge the gap of communication between the different disciplines.

The **Advances in Mechatronics and Mechanical Engineering (AMME) Book Series** provides innovative research and practical developments in the field of mechatronics and mechanical engineering. This series covers a wide variety of application areas in electrical engineering, mechanical engineering, computer and software engineering; essential for academics, practitioners, researchers, and industry leaders.

COVERAGE

- Intelligent Navigation
- Intelligent Sensing
- Vibration and acoustics
- Nanomaterials and nanomanufacturing
- Control Systems Modelling and Analysis
- Autonomous Systems
- Mechanisms and machines
- Medical Robotics
- Micro and nanomechanics
- Computational Mechanics

IGI Global is currently accepting manuscripts for publication within this series. To submit a proposal for a volume in this series, please contact our Acquisition Editors at Acquisitions@igi-global.com or visit: http://www.igi-global.com/publish/.

The Advances in Mechatronics and Mechanical Engineering (AMME) Book Series (ISSN 2328-8205) is published by IGI Global, 701 E. Chocolate Avenue, Hershey, PA 17033-1240, USA, www.igi-global.com. This series is composed of titles available for purchase individually; each title is edited to be contextually exclusive from any other title within the series. For pricing and ordering information please visit http://www.igi-global.com/book-series/advances-mechatronics-mechanical-engineering/73808. Postmaster: Send all address changes to above address.

Titles in this Series

For a list of additional titles in this series, please visit:
http://www.igi-global.com/book-series/advances-mechatronics-mechanical-engineering/73808

Studies on Single and Double Actuator Based DC Attraction Type Levitation Systems Optimization Techniques
Rupam Bhaduri (Dayananda Sagar University, India)
Engineering Science Reference • © 2023 • 231pp • H/C (ISBN: 9781668473887) • US $225.00

Advances in MEMS and Microfluidic Systems
Rajeev Kumar Singh (Department of Mechanical Engineering, Amity University, Noida, India) Rakesh Kumar Phanden (Department of Mechanical Engineering, Amity University, Noida, India) Basant Singh Sikarwar (Department of Mechanical Engineering, Amity University, Noida, India) and J. Paulo Davim (Department of Mechanical Engineering, University of Aveiro, Portugal)
Engineering Science Reference • © 2023 • 331pp • H/C (ISBN: 9781668469521) • US $240.00

Structural Mathematical Modeling Applications in Technological Machines and Transportation Vehicles
Andrey Eliseev (Irkutsk State Transport University, Russia & Irkutsk National Research Technical University, Russia)
Engineering Science Reference • © 2023 • 288pp • H/C (ISBN: 9781668472378) • US $225.00

Handbook of Research on Emerging Designs and Applications for Microwave and Millimeter Wave Circuits
Jamal Zbitou (LABTIC, ENSA of Tangier, University of Abdelmalek Essaadi, Morocco) Mostafa Hefnawi (Royal Military College of Canada, Canada) Fouad Aytouna (LABTIC, ENSA of Tetouan, University of Abdelmalek Essaadi, Morocco) and Ahmed El Oualkadi (LabTIC, ENSA of Tangier, University of Abdelmalek Essaadi, Morocco)
Engineering Science Reference • © 2023 • 542pp • H/C (ISBN: 9781668459553) • US $325.00

For an entire list of titles in this series, please visit:
http://www.igi-global.com/book-series/advances-mechatronics-mechanical-engineering/73808

IGI Global
PUBLISHER of TIMELY KNOWLEDGE

701 East Chocolate Avenue, Hershey, PA 17033, USA
Tel: 717-533-8845 x100 • Fax: 717-533-8661
E-Mail: cust@igi-global.com • www.igi-global.com

Table of Contents

Detailed Table of Contents

The chapter on aerospace image processing with bio-inspired algorithms explores their application to overcome challenges in aerospace image processing. Aerospace and satellite imaging provides valuable insights for environmental monitoring, urban planning, disaster management, and agriculture. This chapter focuses on bio-inspired algorithms, drawing from natural phenomena like genetic evolution and swarm behavior, emphasizing accurate analysis. Various algorithms, including genetic, particle swarm optimization, ant colony optimization, artificial bee colony, firefly, bat, cuckoo search, and artificial neural networks, are discussed in terms of their principles and adaptation for aerospace image processing. Applications in satellite image enhancement, classification, segmentation, and feature extraction are explored. Bio-inspired algorithms are integrated into aerospace image registration, change detection, and image fusion, improving accuracy. Case studies demonstrate their effectiveness in handling large datasets, reducing complexity, and enhancing accuracy.

Different image segmentation algorithms are used for real-time applications like autonomous vehicles, robotics, disaster management, etc. Because of the computational complexity of these algorithms, hardware realizations are cumbersome and complicated. The frames per second achieved are barely sufficient for accurate perception of the problem at hand. The next important challenge is the implementation of evolutionary clustering algorithms like genetic algorithm for improved accuracy, after introducing some simplifying techniques to make the ensuing hardware quicker, less complex with improved power consumption and hardware area/size. Bioinspired algorithms are an excellent candidate solution in this regard. It can be implemented in an FSM based approach to detect faults in the centroid initialization phase like detecting zero data members within a cluster. To further reduce the complexity challenges of existing algorithms while maintaining good accuracy, some new bio-inspired algorithms like roller dung beetle clustering have also been tested.

Blockchain is a special computer system that keeps safe and unchangeable records of transactions and ensures records are real, secure, and many people can access them. The way blockchain works is people share their transactions and others make sure they are real. These transactions become a block which is made secure, users agree, and copies are shared. It is widely used for money transfer, internet of things, etc. Recently, it has been applied in the aerospace engineering sector to store and securely manage aircraft records, using blockchain for better communication in aerospace, to perform predictive analytics in battlefield operations to improve the decision making. Research challenges that are to be addressed while extending to aerospace applications include ensuring security and privacy of information and smart contracts, managing deployment cost, handling huge data traffic generated, and rules and regulations specific to this aerospace sector. In this chapter, blockchain concepts, application to aerospace engineering, recent trends, and future research opportunities are discussed.

A nonlinear control law is designed for quadrotors that considers actuator saturation, external disturbance, and unknown parameter variations that affect the stability of quadrotors. The proposed robust control law is based on sliding mode control with adaptive super twisting algorithm (SMC-ASTA) and has the sliding surface based on proportional derivative (PD) structure of error while the super twisting algorithm inherits proportional integral (PI) properties. Additionally, the discontinuous control gain of SMC-ASTA law is tuned adaptively as functions of sliding surfaces along with an alternative approach of fuzzy logic tuning. The proposed SMC-ASTA demonstrates improved performance in contexts of dynamic stability, command tracking, and control effort compared to the conventional sliding mode control (CSMC). Asymptotic stability of the proposed SMC-ASTA is theoretically justified with Lyapunov-based stability analysis. Comparative analysis of SMC-ASTA in simulation with the conventional SMC demonstrates its superiority and suggests its suitability for practical applications.

The aerospace industry stands at the forefront of technological advancements, seeking innovative solutions to enhance safety, efficiency, and sustainability. This chapter explores the synergies between artificial intelligence (AI) and blockchain technologies, presenting optimization techniques that promise transformative impacts on aerospace engineering practices. The integration of AI algorithms in aerospace operations spans autonomous systems, predictive maintenance, design optimization, and crew support. Autonomous navigation and control benefit from AI's adaptive decision-making, while predictive maintenance algorithms revolutionize aircraft component upkeep, reducing downtime and operational costs. Furthermore, AI-driven design optimization refines aerodynamics and fuel efficiency, ushering in a new era of aircraft performance. This chapter highlights real-world examples that underscore the importance of data integrity in aerospace, showcasing instances where accurate information is critical for mission success.

Industry 5.0 concepts and technologies are increasingly being adopted in the aviation sector, revolutionizing the way aircraft are designed, manufactured, maintained, and operated. Industry 5.0 leverages digital twin technology, creating virtual replicas of physical aircraft, which can be used for real-time monitoring, testing, and design optimization. Engineers and data scientists work together to create and manage these digital twins. Robotics and automation are used for precision manufacturing and assembly processes. Collaborative robots (cobots) are employed to work alongside human technicians, ensuring safer and more efficient assembly.

Satellite images of the entire globe or any given location can be procured quickly with good resolution from any part of the world using thousands of artificial satellites revolving around the Earth. Using advanced AI/ML image processing algorithms, the acquired data can be analyzed to obtain various essential knowledge of any place at any time like the chemical composition of the environment, population density (social), etc. Another prominent field is military, defence, and warfare. Independent hyperspectral image cluster analysis of the world's heavily populated cities like Delhi, Shanghai, etc. has clearly shown the migration of population from rural to the urban. The RS-GIS technology combined with advanced machine learning algorithms predicts that only 50 more years of groundwater supply is left to be harvested. These case observations from different parts of the world show the power and scope of aero-oriented image processing using machine learning algorithms.

Star classification plays a crucial role in understanding the vast and complex universe. Traditional methods of classifying stars are often labour-intensive and time-consuming. The advent of machine learning techniques has opened up new possibilities for automating and improving the accuracy of star classification. This chapter presents an overview of an interstellar star classification system based on machine learning. In this study, the authors propose a novel approach to classify stars using machine-learning algorithms. The dataset comprises a comprehensive collection of stellar data, including spectral characteristics, luminosity, temperature, and other relevant features. The goal is to develop a robust and accurate classification model that can categorize stars into various spectral classes, such as O, B, A, F, G,

K, and M, along with their luminosity classes. This study demonstrates how machine learning could revolutionize the classification of aircraft objects, opening the door to improvements in aeronautical engineering.

Blockchain and artificial intelligence are two emerging technologies with a wide range of industrial uses. Artificial intelligence (AI) has the capacity to learn on its own, which makes it useful for automating some industrial tasks. However, blockchain offers decentralized system features to guarantee auditability, security, and transparency of transactions. The aerospace engineering field can benefit from the integration of blockchain technology and artificial intelligence (AI) to improve productivity, security, and transparency across the board. The area of flight data monitoring is also addressed, with a focus on how blockchain protects vital flight data and AI algorithms examine large datasets to improve flying performance, route optimization, and fuel economy. This chapter further investigates the possibilities of using AI algorithms in conjunction with decentralized air traffic management systems made possible by blockchain technology, which can dynamically adjust to changing circumstances in real time to minimize delays and congestion.

The aerospace industry stands on the brink of transformative advancements with the convergence of artificial intelligence (AI) and blockchain technologies. The chapter commences by highlighting the imperative role of AI and blockchain in the aerospace sector. Subsequently, the chapter delves into the distinct applications of AI within aerospace engineering. Autonomous systems, empowered by AI, are explored for their influence on unmanned aerial vehicles (UAVs), spacecraft navigation, and decision-making. The integration of AI-driven predictive maintenance algorithms emerges as a pivotal strategy to optimize maintenance schedules, reduce downtime, and augment safety through accurate prognostics. Further, the utilization of AI for flight path optimization is detailed, showcasing its ability to reduce fuel consumption and minimize environmental impact. The role of AI extends to the realms of aerodynamics, design, and materials science, accelerating the creation of innovative aircraft and spacecraft designs.

Preface

In the dynamic landscape of Aerospace Engineering, the amalgamation of Artificial Intelligence (AI) and Blockchain technologies has emerged as a transformative force, reshaping traditional paradigms and propelling the field toward unprecedented innovation. This edited reference book, titled *AI and Blockchain Optimization Techniques in Aerospace Engineering*, curated by U. Vignesh, Rahul Ratnakumar, and Abdulkareem Sh. Mahdi Al-Obaidi, encapsulates the cutting-edge advancements, multidisciplinary applications, and novel ideas that define the intersection of AI, Optimization, and Blockchain in aerospace.

The application of Machine Learning, a key facet of AI, has revolutionized the aerospace sector by infusing systems with the ability to learn and adapt from experiences. From air traffic data to flight scheduling information, geographical data, and navigation systems, Machine Learning methods have been proven instrumental in automating complex aerospace processes. Concurrently, Blockchain technology, renowned for its decentralized architecture, has introduced a paradigm shift in ensuring data integrity and security, particularly valuable in an industry where transparency, accountability, and traceability are paramount.

This compendium delves into the synergy of AI and Blockchain, addressing challenges and providing solutions that resonate with the intricacies of aerospace systems. While acknowledging the incredible utility of optimization techniques, the book illuminates unexpected security concerns that have surfaced in recent times. The judicious application of cutting-edge AI and deep learning algorithms, coupled with an extensive array of optimization techniques, emerges as a robust strategy to overcome these challenges.

Each chapter promises a comprehensive exploration of cutting-edge technologies, providing insights for researchers, technologists, and enthusiasts venturing into the forefront of aerospace engineering.

Chapter 1 navigates the application of bio-inspired algorithms in overcoming challenges within aerospace image processing. Emphasizing the importance of accurate analysis in applications such as environmental monitoring and disaster management, the chapter explores various bio-inspired algorithms, including genetic, particle swarm optimization, ant colony optimization, Artificial Bee Colony, Firefly, Bat, Cuckoo Search, and artificial neural networks. The focus extends to satellite image enhancement, classification, segmentation, feature extraction, and applications in aerospace image registration, change detection, and image fusion. Through insightful case studies, the chapter illustrates the effectiveness of these algorithms in handling large datasets, reducing complexity, and enhancing accuracy.

Chapter 2 delves into the challenges of real-time image segmentation algorithms for applications like autonomous vehicles, robotics, and disaster management. Focusing on the computational complexity and hardware realization issues, the chapter introduces bio-inspired algorithms, including Genetic algorithm and Roller Dung beetle clustering, as solutions to enhance accuracy while addressing complexity challenges. The integration of these algorithms in a Finite State Machine (FSM) based approach and their application to fault detection during centroid initialization phase are discussed, presenting a comprehensive approach to reducing complexity challenges in existing algorithms.

Providing a foundational understanding of blockchain, Chapter 3 explores its application in the aerospace engineering sector. Detailing the secure and unchangeable record-keeping nature of blockchain, the chapter discusses its role in storing and managing aircraft records, improving communication in aerospace, and performing predictive analytics in battlefield operations. Key challenges in extending blockchain to aerospace applications, such as security and privacy concerns, deployment cost management, and handling large data traffic, are addressed. The chapter serves as a guide to blockchain concepts, applications, recent trends, and future research opportunities in aerospace engineering.

Chapter 4 focuses on the design of a nonlinear control law for quadrotors, addressing challenges such as actuator saturation, external disturbance, and unknown parameter variations affecting stability. The proposed robust control law, based on sliding mode control with adaptive Super Twisting algorithm (SMC-ASTA), demonstrates superiority over conventional Sliding Mode Control (CSMC). The chapter includes theoretical justification of the proposed SMC-ASTA's asymptotic stability through Lyapunov-based analysis. Comparative simulations showcase improved performance in dynamic stability, command tracking, and control effort, suggesting practical applicability.

Chapter 5 explores the synergies between artificial intelligence (AI) and blockchain technologies in the aerospace industry. Highlighting transformative impacts on safety, efficiency, and sustainability, the chapter covers AI applications in autonomous systems, predictive maintenance, design optimization, and crew support. Real-world examples underscore the importance of data integrity in aerospace, showcasing instances where accurate information is critical for mission success.

Exploring Industry 5.0 concepts in aviation, Chapter 6 delves into the revolutionization of aircraft design, manufacturing, maintenance, and operation. Digital twin technology, robotics, and automation play pivotal roles, creating virtual replicas for real-time monitoring, testing, and design optimization. Collaborative robots (cobots) are introduced for precision manufacturing, ensuring safer and more efficient assembly processes. The chapter sheds light on the collaborative efforts of engineers and data scientists in shaping the future of aviation through Industry 5.0.

Chapter 7 examines the power of AI/ML image processing in analyzing satellite images for global insights. Focusing on applications such as environmental monitoring, disaster mitigation, military and defense, the chapter showcases case observations from different parts of the world. Examples include predicting groundwater supply, tracking population migration, and assessing chemical composition. The integration of remote sensing and geographic information system (RS-GIS) technology with advanced machine learning algorithms highlights the potential of aerospace-oriented image processing.

Presenting a novel approach to star classification, Chapter 8 explores the application of machine learning techniques. The dataset comprises a comprehensive collection of stellar data, including spectral characteristics, luminosity, temperature, and other relevant features. The goal is to develop a robust and accurate classification model categorizing stars into various spectral classes. Beyond astronomy, this study demonstrates the broader implications of machine learning in revolutionizing the classification of objects, opening avenues for advancements in aeronautical engineering.

Chapter 9 explores the integration of blockchain and artificial intelligence in the aerospace engineering field. Addressing productivity, security, and transparency, the chapter discusses applications in flight data monitoring, dynamic air traffic management, and the improvement of flying performance, route optimization, and fuel economy. Challenges such as security, privacy, and regulations specific to the aerospace sector are addressed, emphasizing the potential of blockchain and AI to transform aerospace practices.

The concluding chapter underscores the transformative impacts of the convergence of artificial intelligence (AI) and blockchain technologies in the aerospace industry. Beginning with an overview of the indispensable role of AI and blockchain, the chapter delves into specific applications within aerospace engineering. Topics include autonomous systems, predictive maintenance algorithms, flight path optimization, and advancements in aerodynamics, design, and materials science. Real-world examples highlight the potential of AI and blockchain to shape the future of aerospace engineering.

Our primary objective in compiling this reference book is to provide a platform for researchers to showcase their groundbreaking contributions to the realm of AI and Blockchain applied to Aerospace Engineering and Management. The diverse array of topics covered encompasses technical aeronautical innovations, maintenance strategies, troubleshooting and repair methodologies, control system modeling and design, security frameworks, guidance and navigation systems, predictive fault detection, flight path optimization, and other formidable challenges inherent to the aerospace domain.

Designed for a broad audience, this publication caters to technology developers, academicians, data scientists, industrial professionals, researchers, and students eager to explore the latest innovations in the field. The contributors to this book include academicians, researchers, industrial experts, space scientists, and inspection specialists, fostering a collaborative spirit that reflects the multidisciplinary nature of contemporary aerospace research.

The thematic breadth of this book spans various critical areas, including aerodynamics, flight mechanics, gas dynamics, medical imaging, analytics and diagnostics, image segmentation, computer-aided diagnosis systems, predictive analytics, natural language processing, deep learning methods, speech recognition, text conversation, IoT applications, wearable devices, robotics-assisted systems, pattern recognition, cognitive science, prediction, motif analysis, and bio-inspired algorithms tailored for aerospace applications.

In summation, "AI and Blockchain Optimization Techniques in Aerospace Engineering" stands as a testament to the transformative power of innovative technologies in reshaping the aerospace landscape. We trust that readers, regardless of their background, will find valuable insights and inspiration within these pages, fostering a deeper understanding of the intricate interplay between AI, Optimization, and Blockchain in aerospace systems.

U. Vignesh
Vellore Institute of Technology, Chennai, India

Rahul Ratnakumar
Manipal Institute of Technology, India & Manipal Academy of Higher Education, India

Abdulkareem Sh. Mahdi Al-Obaidi
Taylor's University, Malaysia

Chapter 1
Aerospace Image Processing Using Bio-Inspired Algorithms

K. Ezhilarasan
https://orcid.org/0000-0001-9968-6038
Gandhigram Rural Institute, India

M. Muthukumar
Gandhigram Rural Institute, India

M. Arasumani
University of Greifswald, Germany

A. Jeevarekha
https://orcid.org/0000-0002-5984-5441
M.V. Muthiah Government Arts College for Women, India

S. Karthigai Selvi
https://orcid.org/0000-0001-6249-2037
Galgotias University, India

ABSTRACT

The chapter on aerospace image processing with bio-inspired algorithms explores their application to overcome challenges in aerospace image processing. Aerospace and satellite imaging provides valuable insights for environmental monitoring, urban planning, disaster management, and agriculture. This chapter focuses on bio-inspired algorithms, drawing from natural phenomena like genetic evolution and swarm behavior, emphasizing accurate analysis. Various algorithms, including genetic, particle swarm optimization, ant colony optimization, artificial bee colony, firefly, bat, cuckoo search, and artificial neural networks, are discussed in terms of their principles and adaptation for aerospace image processing. Applications in satellite image enhancement, classification, segmentation, and feature extraction are explored. Bio-inspired algorithms are integrated into aerospace image registration, change detection, and image fusion, improving accuracy. Case studies demonstrate their effectiveness in handling large datasets, reducing complexity, and enhancing accuracy.

DOI: 10.4018/979-8-3693-1491-3.ch001

1. INTRODUCTION

Aerospace image processing is a field that deals with the analysis, enhancement, and interpretation of images captured by satellites orbiting the Earth. These images, acquired using various remote sensing sensors, provide valuable information about the Earth's surface and its dynamic processes. Aerial imagery has proven to be an essential asset in numerous fields, encompassing agriculture, forestry, environmental surveillance, urban planning, disaster management, and beyond. Aerospace imaging and satellite imaging are synonymous.

The field of aerospace image processing has witnessed significant advancements in recent years, thanks to the integration of bio-inspired algorithms. Bio-inspired algorithms draw inspiration from nature's evolutionary and adaptive processes to solve complex problems efficiently. By emulating the principles of natural systems such as genetic evolution, swarm intelligence, and neural networks, these algorithms have demonstrated great potential in satellite image analysis and interpretation.

Aerospace image processing is a specialized area of remote sensing that involves the analysis, manipulation, and interpretation of images acquired by satellites orbiting the Earth. These images, obtained through various remote sensing sensors, provide valuable information about the Earth's surface, atmosphere, and other features from a bird's-eye view.

The primary objective of aerospace image processing is to transform raw satellite data into meaningful and actionable information that can be used for a wide range of applications in diverse fields. It enables researchers, scientists, and decision-makers to gain insights into environmental changes, natural resource management, disaster monitoring, urban planning, agriculture, and more.

2. AEROSPACE IMAGES SOURCES

There are several sources for aerospace images, and availability depends on the purpose and resolution needed. Here are some notable aerospace image sources:

1. NASA Worldview: Provides the capability to interactively browse global, full-resolution satellite imagery and then download the underlying data.
2. Google Earth: Offers high-resolution satellite imagery for various locations around the world. Google Earth Engine also provides access to a large archive of satellite data.
3. USGS Earth Explorer: A vast archive of satellite and aerial imagery, including data from Landsat, Sentinel, and other Earth observation missions.

4. The European Space Agency (ESA) - Copernicus Open Access Hub offers unrestricted access to Sentinel satellite data without any charge.
5. DigitalGlobe (Maxar): A leading commercial provider of high-resolution satellite imagery. They offer services through platforms like SecureWatch and others.
6. Sentinel Hub: Offers access to Sentinel-2 and other satellite data through a simple API.
7. NOAA CLASS, which stands for the National Oceanic and Atmospheric Administration Comprehensive Large Array-data Stewardship System, offers access to diverse satellite datasets.
8. Bing Maps: Microsoft's Bing Maps provides satellite imagery with a bird's-eye view.
9. Planet Explorer: Access to high-frequency satellite imagery from the Planet satellite constellation.
10. Aerial and Satellite Imagery from GIS Companies: Companies like ESRI, Mapbox, and others provide satellite imagery as part of their GIS platforms.

Always be aware of licensing and usage restrictions when accessing and using satellite imagery from these sources.

3. AEROSPACE IMAGE FORMATS

Aerospace images are captured in various formats, each serving specific purposes and applications. Here are some common formats of aerospace images:

1. TIFF (Tagged Image File Format): TIFF is a widely used format for satellite images. It supports lossless compression and can store both raster and vector data. TIFF images are often used for high-quality, uncompressed imagery.
2. JPEG (Joint Photographic Experts Group) is a compressed image format designed for the storage of satellite images. While it offers high compression, it is a lossy format, meaning some image quality is sacrificed for smaller file sizes. JPEG is commonly used for web-based applications.
3. PNG (Portable Network Graphics): PNG is a lossless image format that supports both raster and vector data. It is well-suited for images with transparency. PNG is used when preserving image quality without lossy compression is essential.
4. GeoTIFF: GeoTIFF is an extension of the TIFF format that includes geospatial metadata. It is widely used in GIS (Geographic Information System) applications, allowing for the embedding of coordinate and projection information within the image file.

5. NetCDF (Network Common Data Form): NetCDF is a format used for storing multidimensional scientific data, including satellite imagery. It is commonly used in environmental science and meteorology.
6. HDF (Hierarchical Data Format): HDF is a file format and set of tools for managing complex data. It is often used for storing remote sensing data, including satellite imagery. HDF5 is a version of HDF commonly used for large datasets.
7. KML/KMZ (Keyhole Markup Language) is an XML-based format utilized for presenting geographic data in Earth browsers like Google Earth. KMZ represents a compressed variant of KML, encompassing all essential files within a unified archive.
8. JP2 (JPEG 2000): JPEG 2000 is an improvement over the traditional JPEG format, offering better compression efficiency and quality. It is used for storing high-resolution satellite images.
9. ECW (Enhanced Compressed Wavelet): ECW is a compressed image format designed for efficient storage and rapid display of large raster datasets. It is commonly used in GIS applications.
10. ENVI (Environment for Visualizing Images): ENVI is a popular software-specific format used for remote sensing and image processing. It is commonly used with the ENVI software platform.

These formats serve different needs, from high-quality archival storage to efficient web-based display, depending on the specific requirements of the application.

4. KEY COMPONENTS OF AEROSPACE IMAGE PROCESSING

1. *Image Acquisition:* Satellites equipped with different sensors, such as optical, thermal, and radar sensors, capture electromagnetic radiation from the Earth's surface. The acquired data is transmitted to ground stations for further processing.
2. *Pre-processing:* Raw Aerospace images often contain atmospheric distortions, noise, and other artifacts. Pre-processing involves a series of corrective steps, such as atmospheric correction, radiometric calibration, and geometric rectification, to ensure accurate and georeferenced images.
3. *Image Enhancement:* Image enhancement techniques are applied to improve the visual quality of the images, making them more suitable for human interpretation and analysis. These techniques include contrast stretching, histogram equalization, and filtering.
4. *Image Classification:* Image classification involves categorizing pixels or image segments into different land cover or land use classes. Supervised and

unsupervised classification algorithms are used to automatically assign pixels to specific classes based on their spectral characteristics.

5. *Change Detection:* Change detection techniques compare Aerospace imagesacquired at different times to identify and analyze changes that have occurred on the Earth's surface. This is valuable for monitoring land cover changes, deforestation, urban expansion, and natural disasters.
6. *Fusion ofMultispectral and Hyperspectral Data:* Multispectral and hyperspectral data from different satellite sensors can be combined (fused) to provide more comprehensive and detailed information about the Earth's features and characteristics.
7. *Image Interpretation:* Trained analysts interpret the processed images to extract valuable information relevant to specific applications, such as identifying land cover types, monitoring agricultural crops, or assessing environmental conditions.
8. *Geospatial Analysis:* The processed satellite data is integrated with other geospatial information, such as geographic information system (GIS) data, to conduct spatial analysis and derive meaningful insights.

Advancements in technology, such as cloud computing, machine learning, and artificial intelligence, have significantly influenced the field of satellite image processing. These advancements have enabled faster data processing, automated feature extraction, and the handling of large-scale satellite datasets.

5. BIO-INSPIRED ALGORITHMS

Bio-inspired algorithms, also known as nature-inspired algorithms or evolutionary computation, are computational methods that draw inspiration from natural processes and phenomena to solve complex problems in optimization, search, and decision-making. These algorithms simulate the adaptive and evolutionary mechanisms observed in biological systems, such as genetic evolution, animal behavior, and swarm intelligence.

The idea behind bio-inspired algorithms is to harness the inherent efficiency and adaptability of natural systems to tackle challenging computational problems. By mimicking the principles of evolution, survival of the fittest, and cooperation observed in living organisms, these algorithms have proven to be highly effective in finding optimal solutions to complex and multi-dimensional optimization problems. Some of the major algorithm are given below,

1. *Genetic Algorithms (GA):* Genetic Algorithms (GA) draw inspiration from the principles of natural selection and genetics. They start with an initial population of potential solutions (individuals) and apply genetic operators such as selection, crossover, and mutation to evolve better solutions over generations. The fittest individuals survive and produce offspring, leading to a gradual improvement in the population's overall fitness (Tang, 1996). Ezhilarasan (2024) explained in details about the GA.
2. *Particle Swarm Optimization (PSO):* Particle swarm optimization is inspired by the collective behavior of bird flocks or fish schools. In PSO, a group of particles (representing potential solutions) moves through the solution space, adjusting their positions based on their own experience and the experiences of other particles in the swarm. The swarm cooperates to search for the optimal solution (Abdelbar, 2005).
3. *Ant Colony Optimization (ACO):* Ant colony optimization is based on the foraging behavior of ants. Artificial ants deposit pheromone trails on the paths they explore, and other ants follow these trails to find the shortest path to a food source. In ACO, pheromone trails represent the quality of solutions, and the algorithm converges towards the best solution over time (Dorigo, 1992).
4. *Artificial Neural Networks (ANN):* Inspired by the structure and functioning of the human brain, artificial neural networks are used for machine learning and pattern recognition tasks. ANNs consist of interconnected nodes (neurons) organized in layers, and they learn from data through a process of training and adjusting connection weights (McCulloch, 1943).
5. *Differential Evolution (DE)* is a powerful optimization algorithm that belongs to the family of evolutionary algorithms (Storn, 1997). DE has since gained popularity due to its simplicity, efficiency, and effectiveness in solving complex optimization problems. The algorithm is inspired by the process of natural selection, mimicking the differential reproduction and survival of the fittest individuals in a population. DE operates on a population of candidate solutions and iteratively searches for an optimal solution by using a combination of mutation, crossover, and selection operations. During each iteration, candidate solutions are evolved based on their fitness values and the differences between selected individuals. DE has demonstrated remarkable performance in various domains, including engineering design, financial modeling, and machine learning, making it a valuable tool for global optimization tasks in real-world applications.

The Artificial Bee Colony (ABC)

The Artificial Bee Colony (ABC) algorithm is a bio-inspired optimization technique based on the foraging behavior of honeybees. It was introduced by Derviş Karaboğa in 2005. The ABC algorithm is used to solve various optimization problems, and it can be applied to tasks like image processing, data clustering, and machine learning. The working principle of the ABC algorithm can be summarized in the following steps:

1. Initialization:

Initialize a population of artificial bees (candidate solutions) with random positions in the solution space. Each bee represents a potential solution to the optimization problem.

2. Employed Bees Phase:

In this stage, worker bees navigate the solution space, inspecting various food sources (solutions). Each worker bee assesses the merit of its present solution through an objective function. Information regarding the quality of solutions is exchanged among colony members. Employed bees select a neighbor bee based on the quality of the solutions. They can recruit a bee that performs better than themselves.

3. Onlooker Bees Phase:

Forager bees select a food source for exploration using the insights shared by the working bees. The likelihood of opting for a specific food source is contingent on the excellence of the solution it signifies. Bees associated with superior solutions have a higher probability of being selected as foragers.

4. Scout Bees Phase:

Scout bees oversee the quality of food sources, and if a particular source (solution) shows no improvement over a specified number of iterations (generations), it is discarded. Subsequently, scout bees initiate a random search to find a new food source, replacing the one that has been abandoned.

5. Termination Criteria:

The ABC algorithm continues to iterate through the phases until a termination criterion is met. This could be a maximum number of iterations, a specific solution quality threshold, or other stopping conditions.

6. Solution Update:

As the ABC algorithm progresses, the best solution found by the colony is updated, and it represents the optimal solution to the optimization problem.

7. Optimization Objective:

The ABC algorithm seeks to find the best solution by guiding the artificial bees to explore the solution space, share information, and adapt their behavior based on the quality of solutions.

The ABC algorithm mimics the natural foraging behavior of honeybees, where bees communicate the location of food sources and explore different sources based on their quality. By doing so, it efficiently explores the solution space and converges toward an optimal or near-optimal solution to various optimization problems.

Figure 1. Flowchart of ABC algorithm

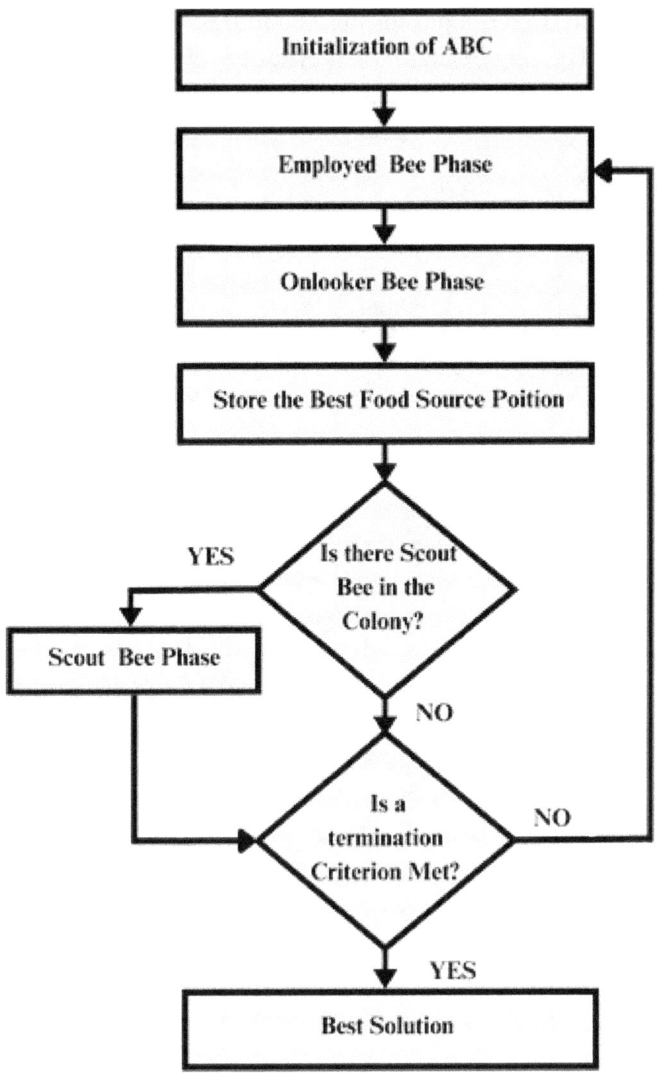

ABC's flexibility and simplicity make it a popular choice for solving a wide range of optimization problems in various fields, including engineering, data mining, and machine learning.

The Firefly Algorithm (FA)

The Firefly Algorithm (FA) is a bio-inspired optimization algorithm that draws its inspiration from the flashing behavior of fireflies to solve optimization problems. It was introduced by Xin-She Yang in 2008. Figure 2 shows life cycle of Firefly.

Figure 2. Shows life cycle of firefly

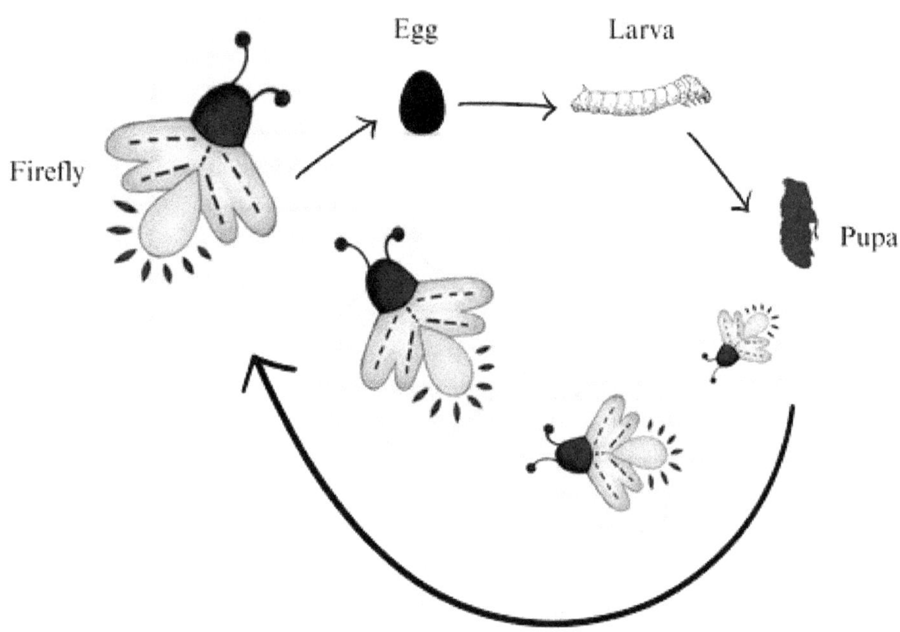

The working principle of the Firefly Algorithm can be summarized in the following steps:

1. Initialization:

Initialize a population of fireflies, each representing a potential solution to the optimization problem. The positions of fireflies are randomly generated in the solution space.

2. Objective Function:

Evaluate the quality of each firefly's solution using an objective function. This function quantifies how close a solution is to the optimal solution, and the goal is to minimize or maximize it, depending on the specific problem.

3. Attraction and Intensity:

Fireflies exhibit an attractive behavior towards other fireflies. The attractiveness of one firefly to another is influenced by their brightness, which is determined by the fitness of their solutions. Brighter fireflies are better solutions. The attraction between two fireflies is inversely proportional to their distance in the solution space. Closer fireflies attract each other more strongly.

4. Firefly Movement:

Each firefly moves towards the brightest (best) fireflies in its vicinity. The movement is guided by the following formula:

New Position = Current Position + β * (brightness of target firefly - brightness of the current firefly) + α * random step size

In this formula, β represents the attractiveness factor, α represents a randomization factor, and the random step size introduces stochasticity in the movement to avoid getting stuck in local optima.

5. Intensity Update:

After each firefly movement, the brightness (fitness) of the fireflies is recalculated based on the objective function.

6. Termination Criteria:

The Firefly Algorithm continues to iterate through the movement and update steps until a termination criterion is met. This could be a maximum number of iterations, a specific solution quality threshold, or other stopping conditions.

7. Solution Update:

As the algorithm progresses, the brightest firefly represents the best solution found by the algorithm.

8. Optimization Objective:

The primary goal of the Firefly Algorithm is to find the best solution by simulating the flashing and attraction behaviors of fireflies. Fireflies with higher brightness (better solutions) are more likely to attract other fireflies, leading to the exploration of the solution space and convergence toward an optimal or near-optimal solution.

The Firefly algorithm is particularly suitable for solving optimization problems with multiple optima or when the problem landscape is complex. Its simplicity, ability to handle multi-modal problems, and global optimization make it a valuable tool in various fields, including engineering, economics, and machine learning. The algorithm's parameter settings, such as β and α, can be adjusted to fine-tune its performance for specific problem domains.

The Bat Algorithm (BA)

The Bat Algorithm (BA) is a bio-inspired optimization technique drawing inspiration from the echolocation behavior of bats. Conceived by Xin-She Yang in 2010, this algorithm is applied for tackling optimization problems.

Figure 3. Shows bat path to reach prey

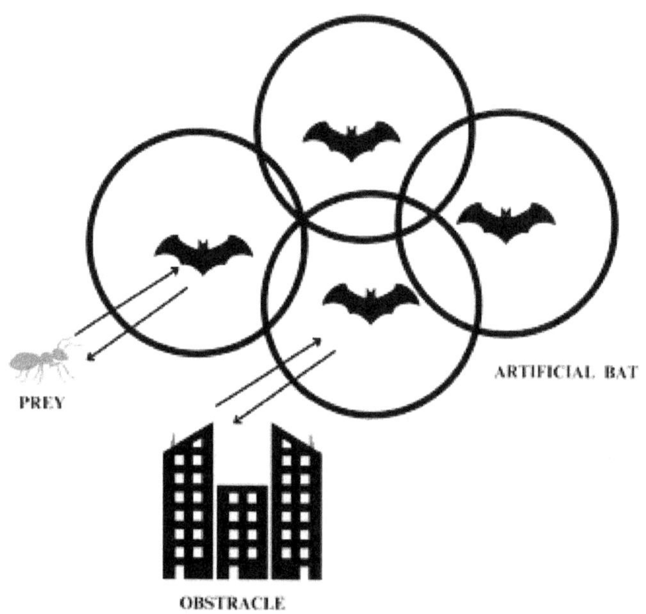

The working principle of the Bat Algorithm can be summarized in the following steps:

1. Initialization:

Initialize a population of bats, each representing a potential solution to the optimization problem. The positions of the bats are randomly generated in the solution space.

2. Objective Function:

Assess the effectiveness of each bat's solution through an objective function, which measures the proximity of a solution to the optimal solution. The objective is to either minimize or maximize this function, depending on the nature of the specific problem.

3. Echolocation and Loudness:

Each bat emits ultrasonic pulses (calls) and listens for echoes to locate prey (optimal solutions). The intensity (loudness) of the bat's calls is related to the quality of its solution, where brighter bats have better solutions. The loudness of a bat decreases as it repeatedly emits calls. The initial loudness and the rate of loudness reduction are user-defined parameters.

4. Frequency and Pulse Rate:

Bats use a frequency tuning mechanism to adjust the wavelength of their calls. The frequency of a bat's calls determines how far it can sense prey. Bats also exhibit pulse rate adjustment, where they control the rate of call emission.

5. Random Walk and Movement:

Each bat performs a random walk in the solution space during each iteration. The random walk is influenced by the bat's loudness and frequency. Bats with higher loudness explore the solution space more extensively. Bats can also adjust their frequency and pulse rate as they search for optimal solutions. The frequency and pulse rate determine how fast a bat explores the space. If a bat finds a better solution than its current one, it moves closer to that solution based on its echolocation behavior. This simulates a guided flight toward better solutions.

6. Solution Update:

As the algorithm progresses, the quality of the solutions held by the bats is updated based on the objective function evaluations.

7. Best Solution:

The best solution encountered by the bats represents the global best solution found by the algorithm.

8. Termination Criteria:

The Bat Algorithm continues to iterate through the random walk and movement steps until a termination criterion is met. This could be a maximum number of iterations, a specific solution quality threshold, or other stopping conditions.

The primary goal of the Bat algorithm is to find the best solution by simulating the echolocation and hunting behavior of bats. Bats with higher loudness (better solutions) are more likely to locate optimal solutions. By exploring the solution space and adjusting their frequency and pulse rate, the bats aim to converge toward an optimal or near-optimal solution. The algorithm's parameters, such as loudness, frequency, and pulse rate, can be adjusted to fine-tune its performance for specific problem domains.

The Cuckoo Search Algorithm (CSA)

The Cuckoo Search Algorithm (CSA) is a bio-inspired optimization approach that draws inspiration from the brood parasitism observed in certain species of cuckoos. Xin-She Yang and Suash Deb introduced this algorithm in 2009 for solving optimization problems

Figure 4. Cuckoo life style

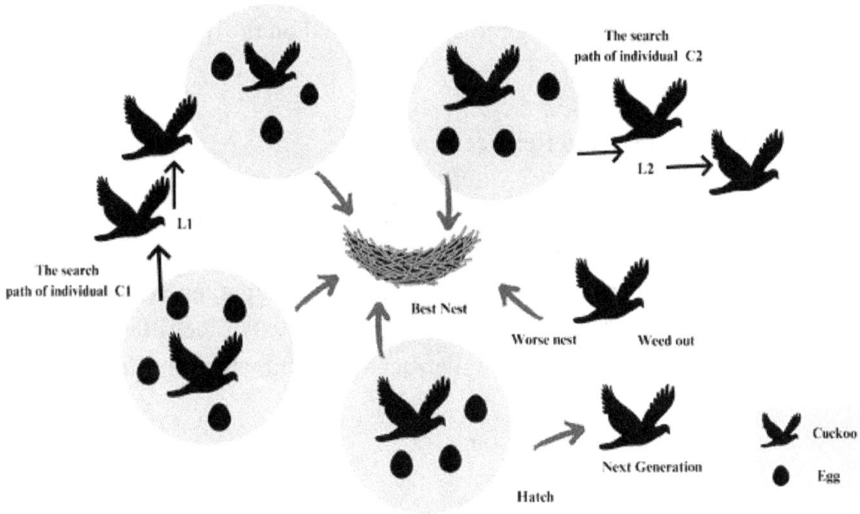

The algorithm is used to find the optimal solution to an optimization problem. Here's an overview of the working principle of the Cuckoo Search Algorithm:

1. Initialization:

Generate a population of host nests, symbolizing potential solutions to the optimization problem. These nests are randomly placed within the solution space. Assess the merit of each nest (solution) using an objective function, which measures the proximity of a solution to the optimal solution. The objective is to either minimize or maximize this function, contingent on the specific nature of the problem.

2. Levy Flights:

The Cuckoo Search Algorithm introduces the concept of "Levy flights," which is a random walk process. Cuckoos lay their eggs in the nests of other bird species, which implies that they might not always choose the best nest. To model this behavior, some nests (solutions) undergo random changes inspired by the Levy distribution. Levy flights are used to explore the solution space, allowing for long jumps and occasionally leading to the discovery of better solutions. The Levy flight steps are generated using a Levy distribution.

3. Cuckoo Nest Selection:

Cuckoos lay their eggs in other birds' nests. In the context of the algorithm, some nests are selected for egg laying, while others are not. The probability of a cuckoo's egg replacing a host's egg in a nest is based on the quality of the cuckoo's solution and the host's nest.

4. Solution Evaluation and Replacement:

After cuckoos have laid their eggs (i.e., proposed new solutions), evaluate the quality of these eggs (solutions) using the objective function. If the quality of a cuckoo's egg is better than the quality of the host nest's egg it replaced, the cuckoo's egg remains in the nest. This process reflects the survival of the fittest solutions. In this way, the population evolves, and higher-quality solutions become more prevalent.

5. Abandonment and Regeneration:

In some cases, cuckoos may abandon their eggs (solutions) when exploring new nests (solutions). This abandonment rate is another user-defined parameter. Abandoned nests (solutions) are replaced by generating new nests randomly in the solution space, promoting exploration.

6. Best Solution:

Track and retain the best solution encountered during the algorithm's execution as the global best solution found.

7. Termination Criteria:

The Cuckoo Search Algorithm continues to iterate, with cuckoos laying eggs, evaluating solutions, and potentially abandoning nests, until a termination criterion is met. This could be a maximum number of iterations, a specific solution quality threshold, or other stopping conditions. The primary goal of the Cuckoo Search Algorithm is to explore the solution space effectively by balancing between exploitation (replacing eggs in nests with better solutions) and exploration (abandoning nests and generating new ones). By applying Levy flights and mimicking the brood parasitism behavior of cuckoos, the algorithm aims to converge toward an optimal or near-optimal solution. User-defined parameters, such as the number of cuckoos, the probability of nest replacement, and the rate of abandonment, can be adjusted to fine-tune the algorithm's performance for specific problem domains.

Bio-inspired algorithms find applications in various fields, including:

- Optimization problems in engineering, logistics, finance, and telecommunications.
- Robotic path planning and control.
- Image and signal processing, feature extraction, and pattern recognition.
- Machine learning, data mining, and predictive modeling.
- Scheduling and resource allocation in manufacturing and transportation.
- Designing efficient and resilient communication networks.
- Financial modeling and portfolio optimization.
- Game playing and decision-making in artificial intelligence.

6. AEROSPACE IMAGE ENHANCEMENT USING BIO-INSPIRED ALGORITHMS

Image enhancement is a crucial step in image processing that aims to improve the quality, clarity, and interpretability of digital images. Traditional image enhancement techniques involve adjusting pixel values to improve brightness, contrast, and other visual aspects. However, bio-inspired algorithms offer a unique and powerful approach to image enhancement by simulating the adaptive and evolutionary processes found in nature. Bio-inspired image enhancement algorithms draw inspiration from natural systems such as genetic evolution, swarm behavior, and neural networks. These algorithms work iteratively, optimizing image properties based on fitness criteria, just as biological systems evolve and adapt to their environments.

Dhal (2015) introduced a novel image enhancement method incorporating a customized version of the Differential Evolution (DE) algorithm with a chaotic Lévy Flight. The study contends that this adapted DE algorithm outperforms conventional DE and DE with a chaotic sequence in terms of both convergence rate and diversity property. Additionally, the paper explores various metaheuristic algorithms successfully employed in image enhancement and segmentation. Emphasis is placed on the significance of image enhancement in augmenting the capability of human or machine recognition systems to extract valuable information from images. The conclusion presents simulation results substantiating the efficacy of the proposed technique in enhancing images.

In 2013, Hoseini (2013) introduced an innovative approach to augment image contrast using a hybrid algorithm that integrates Ant Colony Optimization (ACO), Genetic Algorithm (GA), and Simulated Annealing (SA). This algorithm produces transfer functions mapping input intensities to output intensities, ensuring a harmonious balance between contrast and naturalness. ACO generates transfer functions by considering ant movement directions, with GA automatically determining ant characteristics and directions. SA, functioning as a local search, refines and improves transfer functions and pheromone traces generated by ACO. The fitness criterion proposed eliminates the need for subjective user evaluation and external parameters. Comparative analysis against commonly used techniques reveals the superiority of this approach from both subjective and objective perspectives. The method attains superior uniformity, contrast, DV, and BV values compared to alternative techniques, making it a valuable tool for enhancing images for human observation or as a pre-processing step for various image processing applications.

Qiu (2003) used artificial neural networks (ANNs) for image classification in remote sensing. The article explains the limitations of traditional classification methods and how ANNs can overcome these limitations. It also discusses the architecture of ANNs and how they can be trained using backpropagation. The article presents a case study of using a fuzzy neuro-system for land cover classification in Jacksonville Beach, FL, USA. The results show that the fuzzy neuro-system outperformed other classification methods, including a back-propagation based neural network and a maximum likelihood statistical classification algorithm. The article concludes that ANNs have great potential for image classification in remote sensing, but more research is needed to improve their accuracy and interpretability.

Maulik (2009) proposes a new technique for clustering pixels in satellite images, which can effectively detect regions of varying sizes. The authors view the problem of classifying an image into different homogeneous regions as the task of clustering the pixels in the intensity space. They introduce a modified differential evolution-based fuzzy clustering algorithm that incorporates cluster validity measures and statistical significant tests to improve the accuracy of the clustering results. The proposed technique is compared with traditional fuzzy clustering methods, and the results show that it outperforms them in terms of accuracy and robustness. The article concludes by discussing the potential applications of the proposed technique in remote sensing and other related fields.

In 1999, Tso (1999) investigated the integration of contextual information and random field models to enhance the precision of classification algorithms in the analysis of remotely sensed data. The study illustrates that employing a genetic algorithm results in enhanced parameter estimation and increased classification accuracies. Tso also delves into the theoretical and mathematical underpinnings of Markov random fields and their relevance in handling multisource data. The conclusion drawn is that the inclusion of contextual information effectively enhances the performance of classifiers, emphasizing the necessity for additional research to formulate more intricate models and streamline the parameter estimation process to achieve superior classification accuracy.

In 2013, Basiri (2013) proposed an innovative algorithm grounded in ant colony optimization (ACO) for feature selection in machine learning. The authors posit that feature selection is a pivotal stage in machine learning, contributing to enhanced classification accuracy and diminished computational complexity. They conduct a comprehensive review of existing work on feature selection, delineating the distinctions between model-free and model-based approaches. Introducing their ACO-based algorithm, ACOFS, the method aims to identify a feature set characterized by high classification accuracy and a compact size. Evaluation on both image databases and non-image datasets demonstrates the superior performance of ACOFS compared to other ACO-based feature selection methods and competitive outcomes relative to contemporary state-of-the-art methods. The authors conclude that ACOFS stands out as an efficient and effective approach for feature selection in machine learning.

The advantages of using bio-inspired algorithms for image enhancement lie in their ability to search a vast solution space efficiently and adaptively. They are particularly useful in cases where traditional enhancement methods might not yield satisfactory results or when dealing with complex and noisy images. Satellite image enhancement using bio-inspired algorithms presents a promising approach to improve the visual quality and interpretability of remote sensing data. The integration of bio-inspired algorithms in image enhancement offers robustness, adaptability, and global optimization capabilities. As research in this field continues to evolve, bio-inspired algorithms are expected to play an increasingly significant role in advancing satellite image enhancement techniques, contributing to more accurate and effective analysis in various remote sensing applications.

7. AEROSPACE IMAGE CLASSIFICATION WITH BIO-INSPIRED ALGORITHMS

Image classification is a fundamental task in image processing and computer vision, where the goal is to categorize an input image into predefined classes or categories. Traditional image classification methods often rely on handcrafted features and machine learning algorithms. However, bio-inspired algorithms offer an alternative approach that draws inspiration from natural processes to improve image classification accuracy and robustness.

In 2008, Liu (2008) introduced an innovative approach to classifying remote-sensing images by leveraging ant colony optimization (ACO). The authors contend that conventional statistical classifiers face limitations in addressing complex classification problems due to their stringent assumptions. The proposed method integrates ACO with a discretization technique, resulting in more straightforward rule sets and enhanced accuracy compared to traditional statistical classifiers. Through experiments conducted in the Guangzhou area, the authors observed that the proposed method surpassed traditional statistical classifiers in terms of classification accuracy. The article furnishes a comprehensive depiction of the proposed method, its implementation, and a detailed discussion of the experimental findings.

Hepner (1989) reports an article which explores the potential of artificial neural networks (ANNs) in satellite image processing and compares it to conventional supervised classification. The researchers performed an initial analysis of classification using a minimal training set, juxtaposing the outcomes of ANN back-propagation classification against the traditional supervised maximum likelihood classification. The findings indicated that the neural network yielded a superior land-cover classification compared to the conventional procedure. This research serves as the basis for establishing application parameters, laying the groundwork for subsequent prototyping of software and hardware implementations involving artificial neural networks in satellite image and geographic information processing.

The advantages of using bio-inspired algorithms for image classification lie in their ability to handle high-dimensional feature spaces, adapt to changing environments, and search for better solutions efficiently. Additionally, they can help overcome the curse of dimensionality and enhance the generalization ability of classifiers.

Bio-inspired image classification techniques have been applied to various applications, including object recognition, scene classification, medical image diagnosis, and satellite image analysis. They continue to be an active area of research, and their integration with other machine learning approaches has the potential to further improve the accuracy and robustness of image classification systems. Satellite image classification with bio-inspired algorithms presents a promising approach to effectively utilize remote sensing data for various applications. The integration of bio-inspired algorithms in image classification provides robustness, adaptability, and global optimization capabilities. As research continues to advance in this area, the use of bio-inspired algorithms for satellite image classification is expected to enhance the accuracy, efficiency, and applicability of remote sensing technology, further contributing to sustainable development and informed decision-making in diverse fields.

8. AEROSPACE IMAGE SEGMENTATION USING BIO-INSPIRED ALGORITHMS

Aerospace image segmentation is a critical task in the field of remote sensing and computer vision. It involves dividing a satellite image into distinct regions or segments, where each segment represents a specific land cover or land use class. Accurate and efficient segmentation is essential for various applications, such as urban planning, agricultural monitoring, disaster assessment, and environmental analysis. Traditional segmentation methods often face challenges in handling the complexity and variability of satellite imagery, especially in large-scale and high-resolution datasets. As a result, researchers have turned to bio-inspired algorithms, which have shown promising capabilities in addressing optimization problems and handling diverse data characteristics.

In 2017, Mishra (2017) introduced a methodology for segmenting aerospace images utilizing type 2 fuzzy systems and differential evolution. The proposed algorithm takes into account various considerations inherent in satellite images, such as the Hughes phenomenon and the high correlation among spectral bands, to generate precise and accurate segments. Validation of the algorithm involves comparing cluster validity indexes, specifically the silhouette index, with established algorithms. The paper encompasses a succinct review of prior methodologies, experimental outcomes, and analytical discussions. Given the challenges posed by the uncertainty and high-dimensional space in satellite image segmentation, fuzzy clustering techniques emerge as a suitable approach. The primary objective of satellite image segmentation is to delineate distinct land covers, and the proposed algorithm effectively partitions the image into regions, constituting a comprehensive land cover map.

In 2020, Ezhilarasan and Kiruthika (2020) presented an automated technique for building detection in satellite imagery, employing thresholding and morphological operations. Google Maps images were utilized in their approach. However, the method faced challenges in accurately detecting buildings within town regions.

Awad (2010) describes a new method for satellite image segmentation called TSOM, which uses an unsupervised artificial neural network method called Kohonen's self-organizing map and a threshold technique to cluster Aerospace images into disjoint regions. The paper explains the implementation of SOM and T-Cluster, and provides details on the TSOM segmentation method. The author, Mohamad Awad, conducted experiments on two different Aerospace images to confirm the stability, homogeneity, and efficiency of the TSOM method compared to other methods. The paper concludes that the TSOM method is a suitable option for satellite image processing.

In 2007, Awad (2007) proposed a novel technique for the segmentation of multicomponent images, employing a fusion of a self-organizing map (SOM) and a hybrid genetic algorithm (HGA). The paper underscores the significance of image segmentation in the broader context of image analysis, highlighting that the efficacy of segmentation methods hinges on various factors, including the inherent characteristics of acquired images and the extent of imperfections during the acquisition process. The sequential process involving SOM and HGA is detailed, and a comparative analysis is conducted with the Iterative Self-Organizing DATA Analysis Technique (ISODATA). The conclusion drawn asserts the efficiency and robustness of the proposed method relative to other techniques, positioning it as a viable solution for the segmentation of multicomponent images.

In 2015, Sarkar (2015) introduced an approach for hyper-spectral image segmentation utilizing multi-level thresholding assisted by differential evolution. The method employs Rényi entropy to ascertain optimal threshold values for effective image segmentation. Through experimentation on various hyper-spectral images, the authors conducted comparisons with other state-of-the-art techniques. The outcomes demonstrated the superior performance of the proposed method in both segmentation accuracy and computational efficiency compared to alternative methods. The article furnishes a comprehensive portrayal of the proposed methodology along with detailed insights into the experimental results.

In 2019, Jiang (2019) introduced a highly effective method for satellite image segmentation, leveraging a hybrid grasshopper optimization algorithm (GOA) and minimum cross entropy (MCE). Termed GOA-jDE, this approach amalgamates GOA with self-adaptive differential evolution (jDE) to enhance search efficiency and maintain population diversity. The authors conducted comprehensive experiments using diverse aerospace images, contrasting the proposed method with standard color image thresholding techniques and advanced satellite image thresholding approaches based on various criteria. Results revealed the superior performance of the proposed hybrid algorithm across multiple metrics, including average fitness function value, peak signal-to-noise ratio (PSNR), structural similarity index (SSIM), feature similarity index (FSIM), standard deviation (STD), convergence performance, and computation time. Additionally, the authors validated the robustness of the proposed method using natural images from the Berkeley segmentation dataset. The study's key contributions include the proposition of an efficient satellite image segmentation method, the application of the hybrid GOA and DE algorithm to multilevel thresholding, and the introduction of an alternative hybrid model for a meta-heuristic algorithm.

In 2020, Ezhilarasan and Vennila (2020) introduced an automated approach for shadow detection through the application of morphological operations. Their methodology utilized aerospace images sourced from Google Maps. However, the approach exhibited limitations, particularly in detecting shadows in images with low light conditions.

Bhandari (2015) discusses the challenges of accurate segmentation of remote sensing or Aerospace images due to the high rate of information and rapid rate of change from one region to another. The authors utilize three commonly used image segmentation methods and compare them with the meta-heuristic genetic algorithm. The paper concludes that genetic algorithm is efficient and successful in multilevel thresholding problems, but it requires exponential time. The article provides insights into the application of genetic algorithm for multilevel thresholding in favor of image segmentation.

Satellite image segmentation using bio-inspired algorithms has emerged as a promising research area. The application of bio-inspired algorithms has shown encouraging results in segmenting satellite imagery accurately and efficiently. As the field continues to advance, researchers are exploring hybrid approaches, integrating domain-specific knowledge, and optimizing algorithm parameters to further enhance segmentation performance and address the challenges posed by remote sensing data. The ongoing development of bio-inspired algorithms in satellite image segmentation opens up new possibilities for improved land cover mapping, resource monitoring, and environmental analysis with applications in various domains, including agriculture, urban planning, and disaster management.

9. AEROSPACE IMAGE REGISTRATION USING BIO-INSPIRED ALGORITHMS

Aerospace image registration is a critical task in remote sensing and geospatial analysis, where the objective is to align multiple aerospace images taken at different times or from different sensors onto a common coordinate system. Image registration is essential for various applications, such as change detection, image fusion, and multi-temporal analysis. Traditional image registration methods often rely on feature-based techniques, which may face challenges in handling complex image transformations, occlusions, and sensor variations. To address these challenges, researchers have turned to bio-inspired algorithms, which draw inspiration from nature's adaptive and evolutionary processes, to enhance satellite image registration effectively.

De Falco (2007) presents a parallel software system based on Differential Evolution for the registration of remotely sensed images. The authors describe the design and implementation of the system, and test it on two bi-dimensional remotely sensed images for mosaicking problem. The article explains the basic scheme of Differential Evolution and its application to image registration, as well as the affine transformation and mutual information used in the registration process. The authors also propose a distributed Differential Evolution scheme to improve the efficiency of the system. The results of the testing show that the system is effective in registering remotely sensed images, and has potential applications in the field of remote sensing imagery.

In 2008, De Falco (2008) introduced a software system employing Differential Evolution (DE) for the automatic registration of multi-view and multi-temporal images. The article elucidates the fundamental framework of DE and its application to the registration task. A comparative analysis is conducted against five commonly used registration tools, showcasing the system's effectiveness through experiments on two remote sensing challenges: mosaicking and change discovery. The outcomes illustrate the superior accuracy and efficiency of DE compared to other methods. The article concludes by delving into the potential applications of the system and outlining avenues for future research.

Aerospace image registration using bio-inspired algorithms offers a promising approach to accurately align multiple aerospace images for various remote sensing applications. The integration of bio-inspired algorithms in image registration provides robustness, adaptability, and global optimization capabilities. As research in this field continues to advance, bio-inspired algorithms are expected to play an increasingly significant role in advancing image registration techniques, contributing to more accurate and effective remote sensing analysis and applications.

10. CHANGE DETECTION IN AEROSPACE IMAGE USING BIO-INSPIRED ALGORITHMS

Change detection in aerospace images is a critical task in remote sensing and geospatial analysis, aimed at identifying and quantifying changes that occur between two or more aerospace images taken at different times. These changes could be natural, such as land cover changes due to seasonal variations or human-induced changes like urban expansion or deforestation. Accurate change detection is essential for various applications, including environmental monitoring, disaster assessment, urban planning, and agricultural management. Traditional change detection methods often rely on pixel-based techniques, which may have limitations in handling complex and diverse changes present in satellite images. To address these challenges, researchers have turned to bio-inspired algorithms, which draw inspiration from nature's adaptive and evolutionary processes, to enhance change detection in aerospace images effectively.

In 2010, Celik (2010) introduced an innovative approach for unsupervised change detection in multi-temporal aerospace images, employing a genetic algorithm. The proposed method is designed to yield precise results without relying on prior assumptions. To validate its efficacy, the method is tested on actual multi-temporal datasets representing areas affected by flooding in Bangladesh and parts of India due to prolonged rain, as well as a forest fire in the Reno–Lake Tahoe region of Nevada. The article also delves into the limitations of existing state-of-the-art change detection methods and provides a comparative analysis with the proposed approach.

Mehrotra (2014) presented a research work at conference in the title of an unsupervised change detection technique based on Ant Colony Optimization (ACO) for Landsat 5 satellite images. The proposed method uses ACO to cluster pixels into different classes based on their spectral similarity. The normalized neighborhood ratio approach is used to determine the optimal number of clusters. The method is compared with two other state-of-the-art techniques, PCA k-means and NSCT, and is found to outperform them both visually and quantitatively. The proposed method is shown to be effective in detecting changes in different types of land cover, including vegetation, urban areas, and water bodies. The results demonstrate the potential of ACO-based clustering for unsupervised change detection in satellite imagery.

Change detection in aerospace imagesusing bio-inspired algorithms presents a promising approach to efficiently identify and quantify changes in remote sensing data. The integration of GA, PSO, ACO, and ANN in change detection provides robustness, adaptability, and global optimization capabilities. As research in this field continues to advance, bio-inspired algorithms are expected to play an increasingly significant role in advancing change detection techniques, contributing to more accurate and effective remote sensing analysis and applications.

11. PERFORMANCE EVALUATION AND COMPARATIVE ANALYSIS

Satellite image processing is a critical area of remote sensing and geospatial analysis, aiming to extract valuable information and insights from satellite imagery for various applications. In recent years, bio-inspired algorithms have gained significant attention as powerful optimization techniques in satellite image processing tasks. These algorithms draw inspiration from nature's adaptive and evolutionary processes and offer unique advantages in handling the complexity and variability of satellite images. This chapter focuses on the performance evaluation and comparative analysis of various bio-inspired algorithms applied to different satellite image processing tasks. To provide practical insights, the chapter includes case studies and applications where bio-inspired algorithms are applied to real-world satellite image processing tasks. These applications may include land cover mapping, environmental monitoring, disaster assessment, and urban planning. The case studies demonstrate the effectiveness and versatility of bio-inspired algorithms in addressing diverse challenges in satellite image processing. The performance evaluation and comparative analysis of bio-inspired algorithms in satellite image processing indicate their significant potential in enhancing remote sensing applications. These algorithms offer robust and efficient solutions to complex image processing tasks, outperforming traditional methods in many scenarios. The chapter's findings and case studies contribute valuable knowledge to researchers and practitioners in the field of remote sensing, guiding them in choosing appropriate bio-inspired algorithms for specific satellite image processing tasks and promoting the advancement of satellite image analysis techniques.

When comparing the Artificial Bee Colony (ABC), Firefly Algorithm (FA), Bat Algorithm (BA), and Cuckoo Search Algorithm (CSA) in the context of satellite image processing, certain aspects become more relevant to remote sensing applications. Here's Table 1 shows a comparison focusing on their suitability for aerospace image processing:

Table 1. Comparison of artificial bee colony (ABC), firefly algorithm (FA), bat algorithm (BA), and cuckoo search algorithm (CSA)

Feature	ABC	FA	BA	CSA	Feature
Handling Multidimensional Data	Limited natural adaptation to image data	Limited direct applicability to images	Suitable for image data	Adaptability to image data	Handling Multidimensional Data
Spatial Search and Exploration	Limited direct spatial exploration	Suitable for spatial data exploration	Suitable for spatial data exploration	Applicable to spatial data exploration	Spatial Search and Exploration
Feature Extraction	Limited natural adaptation for features	Limited suitability for feature extraction	Suitable for feature extraction	Adaptability to feature extraction	Feature Extraction
Noise Handling	Limited inherent noise handling ability	Limited inherent noise handling ability	Adaptive mechanisms may aid noise handling	Adaptive mechanisms for noise handling	Noise Handling
Parallel Processing	Limited parallelization potential	Limited parallelization potential	Potential for parallelization	Potential for parallelization	Parallel Processing
Dynamic Data Adaptability	Limited adaptability to dynamic data	Limited adaptability to dynamic data	Adaptive mechanisms for dynamic data	Adaptive mechanisms for dynamic data	Dynamic Data Adaptability
Algorithmic Complexity	Moderate complexity	Moderate complexity	Moderate complexity	Moderate complexity	Algorithmic Complexity
Applications in Image Analysis	Limited for direct pixel-based analysis	Suitable for certain image analysis tasks	Suitable for image analysis tasks	Applicable to various image analysis tasks	Applications in Image Analysis
Performance on Large Datasets	May face challenges with large datasets	May face challenges with large datasets	Capable of handling large datasets	Capable of handling large datasets	Performance on Large Datasets

In the context of aerospace image processing, the suitability of these algorithms depends on the specific tasks and objectives. Some key considerations include their adaptability to spatial data, ability to handle noise, potential for parallelization, and applicability to dynamic datasets. It's essential to choose an algorithm based on the characteristics of the image processing task, the nature of the aerospace data, and the specific goals of the analysis. Additionally, preprocessing steps, such as normalization and feature extraction, play a crucial role in the effectiveness of these algorithms in aerospace image analysis.

12. FUTURE DIRECTIONS AND CHALLENGES

Bio-inspired algorithms have shown great promise in satellite image processing tasks, providing efficient and effective solutions to various challenges. As technology continues to advance and the demand for accurate and timely satellite image analysis grows, the future of bio-inspired algorithms in this field holds exciting possibilities. However, there are also some challenges that need to be addressed to fully leverage the potential of these algorithms for satellite image processing.

Future Directions:

1. *Hybrid Approaches:* One future direction involves exploring hybrid approaches that combine multiple bio-inspired algorithms with traditional image processing techniques. Hybrid methods can leverage the strengths of different algorithms and improve overall performance, leading to more accurate and robust results in satellite image processing tasks.
2. *Deep Learning Integration:* Integrating bio-inspired algorithms with deep learning techniques holds promise for more advanced and automated satellite image analysis. Deep learning can enable the extraction of high-level features and patterns from satellite images, complementing the capabilities of bio-inspired algorithms in handling complex data.
3. *Multi-Sensor Data Fusion:* Future research can focus on developing bio-inspired algorithms specifically tailored for multi-sensor data fusion. Combining data from different satellite sensors can enhance the quality and richness of information extracted from satellite images, enabling more comprehensive analysis.
4. *Real-Time Processing:* Advancements in hardware and parallel computing can enable real-time satellite image processing using bio-inspired algorithms. This will be particularly beneficial for time-sensitive applications such as disaster monitoring and response.
5. *Transfer Learning for Remote Sensing:* Leveraging transfer learning techniques in conjunction with bio-inspired algorithms can improve the efficiency and effectiveness of satellite image processing. Pre-trained models from related

domains can be adapted to remote sensing tasks, reducing the need for large annotated datasets.

Challenges:

1. *Scalability:* As satellite imagery datasets continue to grow in size and complexity, scalability becomes a crucial challenge for bio-inspired algorithms. Ensuring that these algorithms can handle big data efficiently without sacrificing performance will be critical.
2. *Parameter Optimization:* Fine-tuning the parameters of bio-inspired algorithms for different satellite image processing tasks can be time-consuming and computationally expensive. Developing automated methods for parameter optimization is essential to streamline the process and improve algorithm performance.
3. *Robustness to Image Variability:* Aerospace imagescan exhibit significant variability in terms of illumination, resolution, and noise levels. Bio-inspired algorithms need to be robust to handle these variations and produce consistent results across diverse datasets.
4. *Interpretablility:* Interpreting the decisions and results produced by bio-inspired algorithms is important for building trust and understanding their behavior. Enhancing the interpretability of these algorithms in satellite image processing tasks is a significant challenge.
5. *Generalization Across Domains:* Bio-inspired algorithms developed for specific satellite image processing tasks may not generalize well to other domains. Ensuring the transferability of these algorithms across different applications is a challenge that requires careful consideration.

13. CONCLUSION

The future of bio-inspired algorithms in satellite image processing holds immense potential for advancing remote sensing applications. As these algorithms continue to evolve and integrate with other techniques like deep learning, their capabilities are expected to grow exponentially. Addressing the challenges of scalability, parameter optimization, robustness, interpretability, and generalization will be essential to fully exploit the advantages of bio-inspired algorithms in satellite image processing and unlock new possibilities in the field of remote sensing.

In conclusion, bio-inspired algorithms have emerged as powerful and promising optimization techniques in the field of satellite image processing. Drawing inspiration from nature's adaptive and evolutionary processes, these algorithms offer unique advantages in handling the complexity, variability, and high-dimensional nature of satellite imagery. Throughout this chapter, we have explored the applications and capabilities of bio-inspired algorithms in various satellite image processing tasks, including image registration, feature extraction, change detection, image fusion, and classification. The performance evaluation and comparative analysis of bio-inspired algorithms have demonstrated their effectiveness in achieving accurate and consistent results across diverse satellite image datasets. These algorithms exhibit robustness to noise and variations in the data, making them well-suited for real-world satellite imagery. Additionally, their adaptability to different scenarios and image characteristics highlights their potential for enhancing remote sensing applications.

Looking to the future, bio-inspired algorithms hold exciting directions for further advancement in satellite image processing. Hybrid approaches, integration with deep learning techniques, multi-sensor data fusion, and real-time processing are some of the areas that researchers can explore to enhance algorithm performance and efficiency. Moreover, leveraging transfer learning and addressing challenges such as scalability, parameter optimization, robustness, interpretability, and generalization will be crucial in realizing the full potential of these algorithms in satellite image analysis.

In summary, bio-inspired algorithms offer a promising avenue for pushing the boundaries of satellite image processing, enabling accurate and timely analysis of remote sensing data. As technology continues to evolve, these algorithms are expected to play an increasingly significant role in revolutionizing the field of remote sensing, contributing to better environmental monitoring, disaster assessment, land cover mapping, and urban planning. Continued research and development in this area will undoubtedly pave the way for innovative and transformative solutions in satellite image processing.

REFERENCES

Abdelbar, A. M., Abdelshahid, S., & Wunsch, D. C. I. (2005). Fuzzy PSO: A generalization of particle swarm optimization. *Proceedings of 2005 IEEE International Joint Conference on Neural Networks (IJCNN '05),* 1086–1091. 10.1109/IJCNN.2005.1556004

Awad, M. (2010). An unsupervised artificial neural network method for satellite image segmentation. *International Arab Journal of Information Technology, 7*(2), 199 – 205.

Awad, M., Chehdi, K., & Nasri, A. (2007). Multicomponent Image Segmentation Using a Genetic Algorithm and Artificial Neural Network. *IEEE Geoscience and Remote Sensing Letters*, *4*(4), 571–575. doi:10.1109/LGRS.2007.903064

Basiri, M. E., Ghasem-Aghaee, N., & Aghdam, M. H. (2013). Efficient ant colony optimization for image feature selection. *Signal Processing*, *93*(1), 1–14. doi:10.1016/j.sigpro.2012.06.019

Bhandari, A. K., Pare, S., Kumar, A., Singh, S. K., & Kumar, A. (2015). Satellite image segmentation based on different objective functions using genetic algorithm: A comparative study. In *2015 International Conference on Digital Signal Processing* (pp. 1245-1250). IEEE. 10.1109/ICDSP.2015.7251972

Celik, T. (2010). Change detection in satellite images using a genetic algorithm approach. *IEEE Geoscience and Remote Sensing Letters*, *7*(2), 388–392. doi:10.1109/LGRS.2009.2037024

De Falco, I., Della Cioppa, A., Maisto, D., & Tarantino, E. (2008). Differential Evolution as a viable tool for satellite image registration. *Applied Soft Computing, 8*(4), 1453-1462. doi:10.1016/j.asoc.2007.10.013

De Falco, I., Maisto, D., Scafuri, U., Tarantino, E., & Della Cioppa, A. (2007). Distributed Differential Evolution for the Registration of Remotely Sensed Images. In *15th EUROMICRO International Conference on Parallel, Distributed and Network-Based Processing (PDP'07),* Napoli, Italy (pp. 358-362). 10.1109/PDP.2007.36

Dhal, K. G., Quraishi, M. I., & Das, S. (2015). Parameterized image enhancement technique using modified differential evolution via chaotic Lévy flight. *International Journal of Applied Metaheuristic Computing*, *6*(3), 69–81. doi:10.4018/ijamc.2015070104

Dorigo, M. (1992). *Optimization, learning and natural algorithms* [PhD thesis]. Politecnico di Milano, Italy.

Ezhilarasan, K., & Kiruthika, K. (2020). Automatic Building Detection Using Thresholding and Morphological Operations. In *National Conference on Application of Geospatial Technologies in Management of Natural Resources and Disaster (UGC Sponsored) - (AGTMNRD-2020)* (pp. 138-141). Academic Press.

Ezhilarasan, K., Somasundaram, K., Kalaiselvi, T., Praveenkumar, S., Karthigai Selvi, S., & Jeevarekha, A. (2024). Bio-Inspired Algorithms Used in Medical Image Processing. In U. Vignesh, M. Manikandan, & R. Doshi (Eds.), *Bio-Inspired Optimization Techniques in Blockchain Systems* (pp. 1–300). IGI Global. doi:10.4018/979-8-3693-1131-8

Ezhilarasan, K., & Vennila, M. (2020). An automatic shadow detection using mathematical morphological operations. In *National Conference on Application of Geospatial Technologies in Management of Natural Resources and Disaster (UGC Sponsored) - (AGTMNRD-2020)* (pp. 134-137). Academic Press.

Hepner, G., & Ritter, N. (1989). *Application of an artificial neural network to land cover classification of thematic mapper imagery*. JPL Internal Technical Report.

Hoseini, P., & Shayesteh, M. G. (2013). Efficient contrast enhancement of images using hybrid ant colony optimisation, genetic algorithm, and simulated annealing. *Digital Signal Processing*, *23*(3), 879–893. doi:10.1016/j.dsp.2012.12.011

Jiang, H., Li, C., Oliva, D., & Peng, X. (2019). Hybrid grasshopper optimization algorithm and differential evolution for multilevel satellite image segmentation. *Remote Sensing (Basel)*, *11*(9), 1134. doi:10.3390/rs11091134

Liu, X., Liu, L., & Zhang, J. (2008). Ant colony optimization for classification of remote-sensing imagery. *IEEE Transactions on Geoscience and Remote Sensing*, *46*(12), 4190–4199. doi:10.1109/TGRS.2008.2001754

Maulik, U., & Saha, I. (2009). Modified differential evolution based fuzzy clustering for pixel classification in remote sensing imagery. *Pattern Recognition*, *42*(9), 2135–2149. doi:10.1016/j.patcog.2009.01.011

McCulloch, W. S., & Pitts, W. (1943). A logical calculus of the ideas immanent in nervous activity. *The Bulletin of Mathematical Biophysics*, *5*(4), 115–133. doi:10.1007/BF02478259

Mehrotra, A., Singh, K. K., & Singh, P. K. (2014). Unsupervised change detection in satellite images using ant colony optimization clustering. In *2014 International Conference on Computing for Sustainable Global Development (INDIACom)* (pp. 410-415). IEEE. 10.1109/IndiaCom.2014.6828169

Mishra, S., & Singh, S. (2017). Satellite image segmentation based on differential evolution. *Proceedings of the International Conference on Intelligent Sustainable Systems (ICISS 2017)*, 623. https://doi.org/10.1109/ICISS.2017.8388766

Qiu, F., & Jensen, J. R. (2003). Building classification models with combined neural networks for high-resolution satellite imagery. *International Journal of Remote Sensing*, *24*(23), 4551–4570. doi:10.1080/0143116031000116414

Sarkar, S., Das, S., & Chaudhuri, S. S. (2015). Hyper-Spectral Image Segmentation using Rényi Entropy based Multi-Level Thresholding Aided with Differential Evolution. *Expert Systems with Applications*, *42*(7), 3606–3618. doi:10.1016/j.eswa.2015.01.035

Storn, R., & Price, K. (1997). Differential evolution - a simple and efficient heuristic for global optimization over continuous spaces. *Journal of Global Optimization*, *11*(4), 341–359. doi:10.1023/A:1008202821328

Tang, K. S., Man, K. F., Kwong, S., & He, Q. (1996). Genetic algorithms and their applications. *IEEE Signal Processing Magazine*, *13*(6), 22–37. doi:10.1109/79.543973

Tso, B., & Mather, P. M. (1999). Classification of multisource remote sensing data using a genetic algorithm and Markov random fields. *IEEE Transactions on Geoscience and Remote Sensing*, *37*(3), 1367–1377. doi:10.1109/36.763284

Chapter 2
An Empirical Review on Clustering Algorithms for Image Segmentation of Satellite Images

U. Vignesh
Vellore Institute of Technology, Chennai, India

Rahul Ratnakumar
Manipal Institute of Technology, India & Manipal Academy of Higher Education, India

ABSTRACT

Different image segmentation algorithms are used for real-time applications like autonomous vehicles, robotics, disaster management, etc. Because of the computational complexity of these algorithms, hardware realizations are cumbersome and complicated. The frames per second achieved are barely sufficient for accurate perception of the problem at hand. The next important challenge is the implementation of evolutionary clustering algorithms like genetic algorithm for improved accuracy, after introducing some simplifying techniques to make the ensuing hardware quicker, less complex with improved power consumption and hardware area/size. Bioinspired algorithms are an excellent candidate solution in this regard. It can be implemented in an FSM based approach to detect faults in the centroid initialization phase like detecting zero data members within a cluster. To further reduce the complexity challenges of existing algorithms while maintaining good accuracy, some new bio-inspired algorithms like roller dung beetle clustering have also been tested.

DOI: 10.4018/979-8-3693-1491-3.ch002

1. INTRODUCTION

Optimization methods, AI, Machine Learning and Blockchain Techniques are heavily applied in Aerospace Engineering. Machine Learning offers the capability of improving from experience and employing novel self/reinforced-learning methods which can automate many aerospace systems and applications. Machine learning methods have been successfully applied to provide us with valuable insights from air traffic data, flight scheduling information, geographical information, navigation and path planning data, etc. Despite the incredible utility of various optimization techniques, various unexpected security problems have been demonstrated in recent years. Several challenges and multidimensional problems faced by aero-systems can be resolved by the judicious application of cutting-edge AI machine and deep learning algorithms combined with a vast arsenal of optimization techniques.

Nature from time immemorial has been an inspiration for man. She has inspired us to mimic her ways in many meta-heuristic techniques, algorithms and architectures, especially while designing engineering solutions. Operations performed by these are simple but powerful and remarkable ideas can dramatically improve the performance, efficiency, economy and accuracy of the system. Science and Engineering owes a lot to nature. From the Newton's gravitational law to the design of modern buildings, nature has been nourishing human brain with infinite supply of ideas, worth mimicking. Every time Nature comes up with great inspirations and revelations which can revolutionize the technology dramatically of that period. The best examples are airplanes inspired from birds, ships and submarines inspired from aquatic organisms, camera system inspired by human eye, humanoid robots to… Velcro to geko, the list is endless. Even the vast array of Bio-Inspired algorithms and Nature-inspired architectures from- Fibonacci pattern seen in the flowers and shells, to the billions of the connections of the Neural networks termed as Deep learning architectures are nothing but gifts of nature, envisioning those people who tries to approach her from time to time. The profession of Engineering demands a lot of problem solving along with achieving optimal outcomes within a limited period, that too in an economical way. So, Engineers, at times are forced to look outside 'to see and learn how the nature solves the similar problem and how he can replicate the same idea in their designs.' The field of engineering called as biomimetics (biomimicry) implying designs where engineering solutions mimics Nature is growing very rapidly, thanks to Nature.

Satellite images, once cryptic mosaics of pixels, are now yielding their secrets to the power of clustering algorithms in aerospace engineering. These algorithms, akin to celestial cartographers, group similar pixels together, carving the image into meaningful segments, each revealing a piece of the cosmic puzzle. Imagine a vast expanse of ocean captured by a satellite camera. Traditional analysis might struggle to distinguish subtle variations in water color, potentially missing crucial details about currents, phytoplankton blooms, or pollution. But clustering algorithms step in, identifying clusters of pixels with similar spectral signatures. These clusters could then reveal hidden eddies, track the drift of pollutants, or even map the distribution of marine life. Beyond Earth's watery embrace, clustering algorithms tackle the Martian landscape with similar finesse. By grouping pixels based on their brightness, texture, and temperature, they can delineate rock formations, unearth potential ice deposits, and even identify signs of ancient riverbeds. This information is invaluable for planning future robotic missions and piecing together the Red Planet's geological history. The applications extend beyond planetary exploration. Clustering algorithms play a crucial role in Earth observation, segmenting satellite images to distinguish urban sprawl from natural vegetation, monitor deforestation, and track the movement of glaciers. This information empowers policymakers to make informed decisions about land management (Sasaki et al., 2007), environmental protection (Cao et al., 2018), and disaster preparedness (Amit et al., 2017).

The future of satellite image segmentation in aerospace engineering is brimming with possibilities. As algorithms become more sophisticated and incorporate advanced machine learning techniques, they will not only segment images with greater accuracy but also extract deeper insights. Imagine clustering algorithms not just identifying landforms but also recognizing signs of volcanic activity, predicting crop yields, or even monitoring the health of coral reefs. The sky is no longer the limit for satellite image segmentation in aerospace engineering. With clustering algorithms as our celestial guides, we are deciphering the secrets of the cosmos, pixel by pixel, unlocking a universe of knowledge that shapes our understanding of our place in the vast expanse of space.

When it comes to image segmentation architectures also, Nature has always been the source of infinite ideas to efficiently and effectively implement solutions, techniques etc. supporting unparalleled advancements and Innovations. In many real-time image segmentation applications like disaster (fire, earth quake, flood etc.) detection (Ignatiev et al., 2018, Amit et al., 2017) and object tracking (Military, satellite and navigation applications), performance and efficiency are more important than accuracy. Moreover, in other applications like Multimedia, finance and Biomedical engineering economy plays an important role than accuracy.

The following important considerations are derived with respect to hardware design for Machine learning or conventional optimization algorithms for satellite image segmentation

1. Flexibility through Hardware: Modern reconfigurable architectures let us tailor algorithms for hardware, allowing us to sagaciously (approximate computing) remove latencies, complexities, and redundancies during hardware

implementation without sacrificing accuracy. This not only simplifies the hardware but also boosts its performance.

2. Segmentation Everywhere: Segmentation (He, Kaiming et al., 2015) is a crucial step in booming fields like multispectral/hyperspectral imaging (Fujita et al., 2017, (He et al., 2015), satellite imaging, biomedical analysis, biometrics, high-quality audio/video/image analysis, and many real-time multimedia applications.
3. Satellites Demand High-Performance Computing: High-resolution satellite data captured by advanced sensors requires high-performance computing (HPC) platforms due to its enhanced spatial, spectral, and temporal resolutions.
4. Existing Solutions Struggle: Current homogeneous and heterogeneous computing solutions struggle with efficient real-time processing of huge volumes of high-speed satellite image data.
5. Need for Adaptive Hardware: Therefore, there's a critical need for highly flexible, lightweight, and low-power integrated image processing modules that can adapt to on-board remote sensing hardware configurations. These modules would be embedded in the processing units of satellites or space stations.
    5. Need for Intelligent Hardware: This is an advanced stage of adaptability where system predicts its own improvement/ decay based on past experience/ training.
    6. FPGAs: The All-rounder: FPGAs offer several advantages. They boast significantly faster development times than ASICs, achieve performance close to GPUs while consuming less power, and remain flexible for remote real-time programming and upgrades. Additionally, FPGAs exhibit superior tolerance to space radiation and thermal variations compared to other VLSI options. These unique features make them exceptionally promising for space applications.

Recently, a novel algorithm and architecture inspired from the Roller dung beetles, for performing high speed image segmentation, requiring less computational resources with simple architecture, at the cost of accuracy has been proposed by Ratnakumar et al. (2021). This algorithm comprises of only three processes i.e. Merge sorting, data comparison and data labelling without using any mathematical operations. The performance of the algorithm when tested on natural and synthetic images procured from benchmark databases including Landsat Image Gallery of NASA, is found to be excellent considering the parameters like execution time, parallelism and resource utilization, but at the cost of accuracy.

2. HARDWARE ARCHITECTURE OF CLUSTERING FOR IMAGE SEGMENTATION

Real-time image segmentation is essential for undertaking prompt action against natural or manmade calamities like floods, forest-fires, earthquakes (Fig. 1) etc. Nevertheless, due to the lack of dedicated hardware, huge time delay for required processing, limited number of technically trained professionals who have prior knowledge about the possible outcomes, the existing algorithms and techniques have not been successfully utilized for the benefit of thc human race. Existing image segmentation algorithms are computationally complex (like calculus-based methods) and their expensive hardware implementation lacks the speed (active shape and graph-based models), efficiency, flexibility (Mean-shift Algorithm/ Mumford-Shah frameworks) and needs considerable chip area for their real-time and on-line deployment (like Deep learning models). Realtime image segmentation is an unsupervised optimization problem where the given set of unlabelled pixels is grouped based on their similar pixel intensities, within a short time frame. One of the simple and straight forward methods to perform image segmentation (Duarte et al., 2018) is to formulate it as a clustering problem. The major works referred in this chapter is focused on proposing novel reconfigurable hardware platforms of clustering algorithms like Roller Dung Beetle Clustering; Histogram based Automatic K-Means (HISTAK), lower complexity implementation of K-Means (Fig. 2), Genetic Clustering, and Artificial Immune System (AIS) clustering for performing real-time image segmentation (Doshi, 2018). These algorithms need to process billions of image pixels captured in real-time, requiring infinite parallelism! The rise of Field Programmable Gate Arrays (FPGA) with their many advantages like massive parallelism, versatility, flexibility and a short time to market has opened us to a new way to solve this problem. One significant contribution of this empirical review includes implementing a K-Means clustering architecture with a lower computational requirement due to the introduction of simplified Convergence Checker and Fibonacci linear feedback shift register (LFSR) for centroid initialization. The hardware complexity has been further reduced by using Manhattan as the distance metric, the design optimized for power and chip area. Eight latest satellite images of natural and human devastations, taken from NASA's OLI, were successfully segmented, yielding satisfactory PSNR and SSIM values. In order to tackle the problem of finding 'K'-the number of possible clusters within the image, a simplified automatic segmentation technique (Rahman et al., 2016) and its hardware architecture suited for satellite and histopathological image segmentation called HISTAK-'Histogram based Automatic K-Means' combining the K-Means algorithm and a novel statistical analysis technique HIST (Histogram Separation Technique), has been successfully implemented (Ratnakumar et al., 2023). Biopsy images of Cancer, COVID-19, Pneumatic infections have been perfectly segmented. Using this architecture, various distinct features like tumors, multinucleated-cells, cytoplasm, and pulmonary embolisms were segmented in real-time to identify the stage of diseases like Lung cancer, COVID-19 etc. (Ignatiev et al., 2018).

Figure 1. The NASA's Operational Land Imager (OLI) on Landsat 8 captured this image of the North India depicting the flooded area

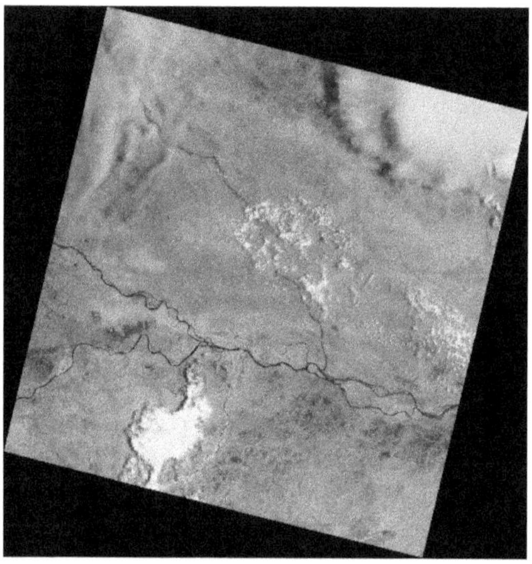

Figure 2. Proposed methodology for satellite image segmentation

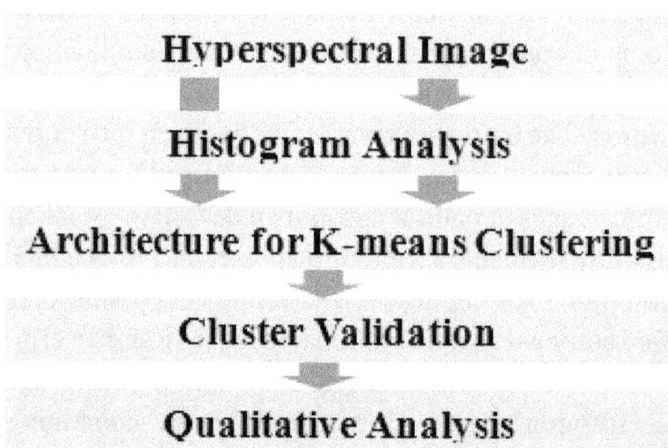

Figure 3. Block diagram of proposed k-means architecture

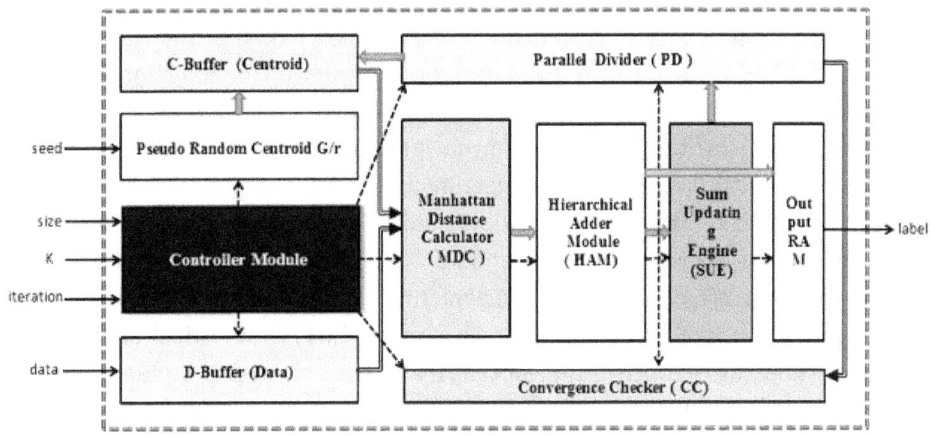

The next important advancement is the implementation of evolutionary clustering architectures like Genetic algorithm and AIS clustering, for improved accuracy, after introducing innovative architectures for crossover and mutation modules and also reducing the number of Pseudo Random Number Generator (P-RNG) units (Fig. 2.). The proposed architecture (Fig. 3.) has been tested on six satellite flood images, with an improved power consumption of 31mW and a clock frequency of 191MHz (Ratnakumar et al., 2019).

Another challenging requirement in hardware architectures is the ability to withstand faults both software and hardware faults, especially in critical real-time image segmentation applications (Duarte et al., 2018). A fault-tolerant FSM architecture based on Artificial Immune system (AIS) is proposed for real-time segmentation (Rahman et al., 2016), where a few redundant states with a high probability of occurrence are introduced within the state machine to cover all the major faults. It can detect faults in the centroid initialization phase like- detecting zero data members within a cluster, detecting a coinciding centroid or centroid placed far beyond data space etc. This helps in an undeterred hardwired execution of the clustering process. Six satellite images (Demir et al., 2018) depicting devastating floods has been successfully segmented with satisfactory PSNR and SSIM values, along with an improved power consumption of 74mW and a very good clock frequency of 217MHz. When the flood-water cluster from the image was compared with the same of a previous image (Ignatiev et al., 2018), it revealed significant insights like percentage increase in the water-cover over the specific digitally mapped regions (Voigt et al., 2016), which could have speed up the rescue efforts and disaster management strategies (Sasaki et al., 2007), involving opening and closing of dams, reservoirs etc.

To further reduce the complexity challenges of existing algorithms while maintaining good accuracy a bio-inspired algorithm based on the behaviour of 'Roller Dung Beetles (RDB)' has been proposed (Ratnakumar et al., 2021), which uses a comparison free sorting algorithm and thresholding-labelling engine. The beauty of this proposed RDB Clustering architecture is its lower computational complexity O(N). It is 42% faster than the conventional architectures with a clock frequency of 230.52MHz with an increased PSNR of 6.825% and SSIM of 15.55%, when applied on satellite images (Demir et al., 2018) depicting natural or manmade disasters like floods, forest fire, ice shrinkage etc. (Amit et al., 2017). Statistical analysis and silhouette index also confirm the superiority of new clustering architecture over existing implementations including CNN (Fig. 4, Fig. 5 and Fig. 6), in terms of lower chip-area and power consumption.

Overall, the review work presented focuses on understanding and proposing a novel reconfigurable hardware platforms of clustering algorithms for executing real-time image segmentation applications (Doshi, 2018), where the speed and computational complexity of the algorithm play a vital role along with accuracy. This review work act as a ray of light for the existing hardware systems and processing tools within satellites, gadgets, UAVs, Drones and other medical diagnosis systems for manifesting 'true intelligence', which is the need of this century.

Figure 4. Clustering output using synthetic image dataset

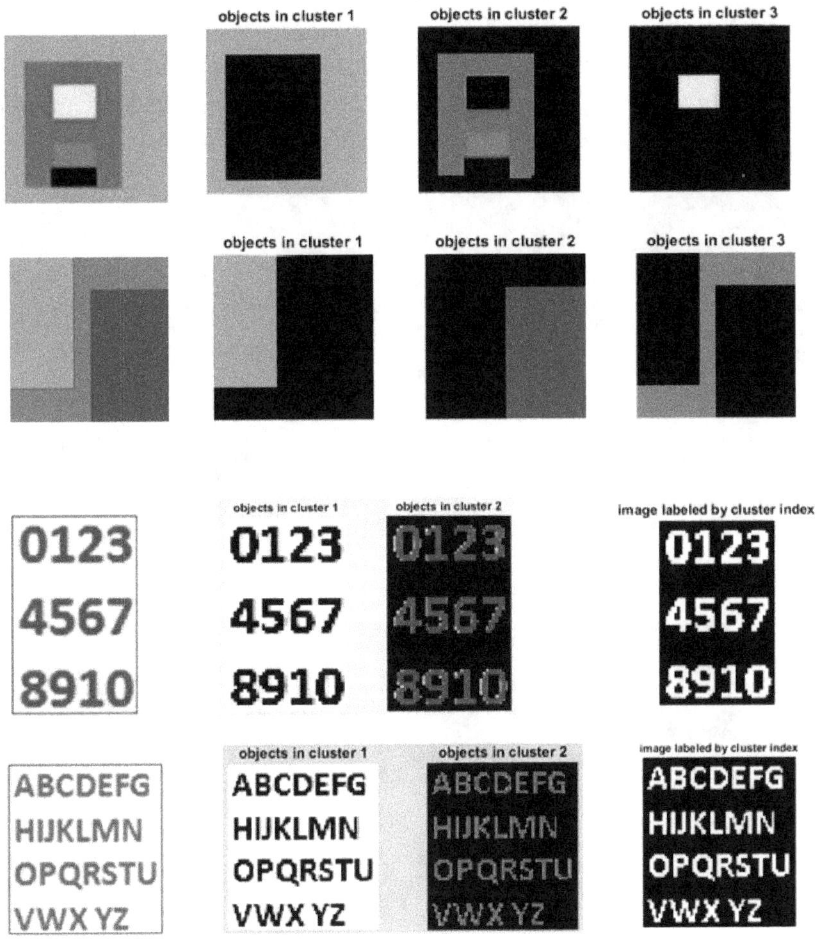

Figure 5. The NASA's Operational Land Imager (OLI) on Landsat 8 captured hyperspectral images related to smoke and clouds (a) image showing the cedarfire at Eastern Arizona on 19th Jun. 2016, (b) image of pyrocumulus clouds and smoke rising from the fire at the south coast of Western Australia on 7th Dec. 2016, (c) image showing the eruption at Holouhraun, Iceland on 6th Sep. 2014, (d) image of the cloud filled canyons of Peru captured on 26th Jul. 2015

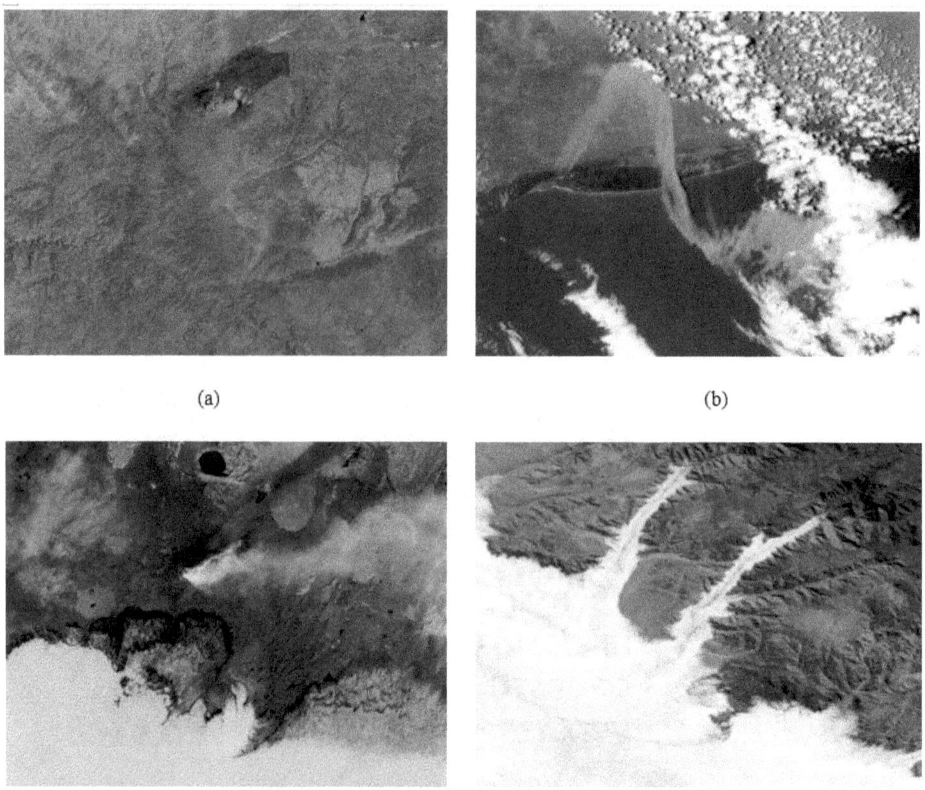

(a) (b) (c) (d)

Figure 6. The NASA's Operational Land Imager (OLI) on Landsat 8 captured hyperspectral images related to water resources (a) image depicting the draught conditions of the lake Cachuma on 8th Nov. 2016, (b) image of the Apostle islands on 14th Aug. 2015, (c) image of the Sierra de sangra glacier on 14th Jan. 2015, revealing the ice shrinkage, (d) image of Kaikoura, New Zealand on 13th Nov. 2016 after the 7.8 magnitude earthquake

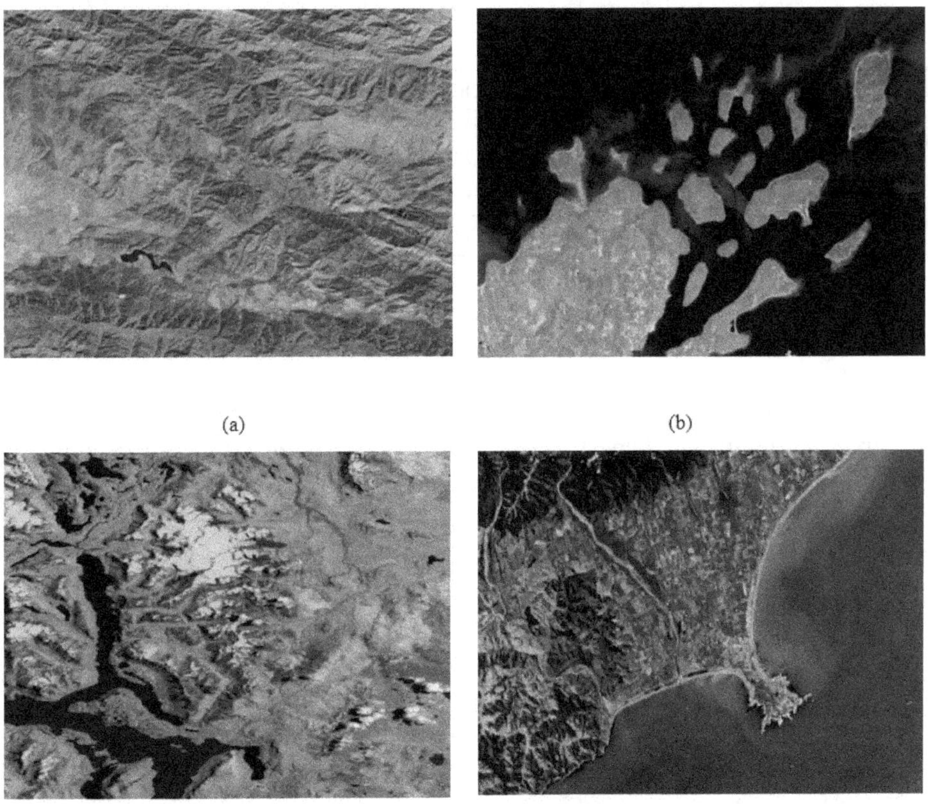

(a) (b)

(c) (d)

Table 1. Specifications of the satellite images used for simulations

Sl	Image	Bit Depth	Sensor/Satellite	Dimension	Date of Capture
1	Cedarfire	32	Landsat 8-OLI	2000 X 2000	Jun 19, 2016
2	Peru_oli	32	Landsat 8-OLI	2899 X 3306	Dec 22, 2016
3	Cocklebiddy_oli	32	Aqua-MODIS Landsat 8 -OLI	2000 X 6000	Dec 7, 2007
4	Apostle_oli	32	Landsat 8-OLI	3900 X 2600	Aug 26, 2016
5	Cachuma_oli	32	Landsat 8-OLI	1090 X 1090	Nov 15, 2016
6	Sierradesangra	32	Landsat 5-TM Landsat 8-OLI	2000 X 3000	Feb 18, 2016
7	Kaikoura	32	Sentinel-2, ESA	1663 X 1021	Nov 13, 2016
8	Holuhraun_oli	32	Landsat 8-OLI	2400 X 1800	Sept 8, 2014

3. EXPERIMENTAL RESULTS AND DISCUSSIONS

Figure 7. Grouping the hyperspectral images into two clusters: (a) Cachuma, (b) Apostle, (c) Sierra de sangra, (d) Kaikoura, (e) Cedarfire, (f) Cocklebiddy, (g) Holuhraun, (h) Peru

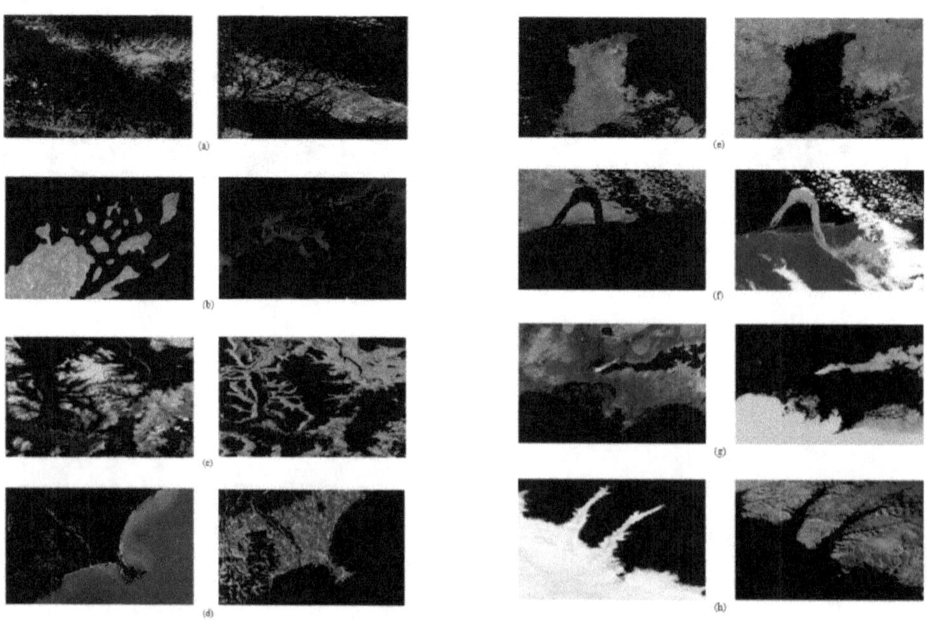

Figure 8. Grouping the hyperspectral images into three clusters: (a) Cachuma, (b) Apostle, (c) Sierra de sangra, (d) Kaikoura, (e) Cedarfire, (f) Cocklebiddy, (g) Holuhraun, (h) Peru

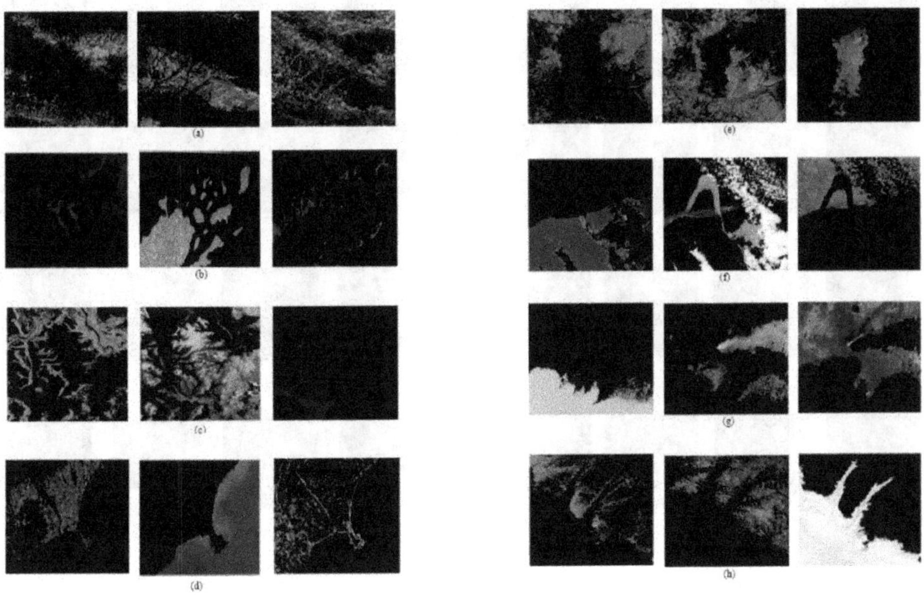

Figure 9. Grouping the hyperspectral images into four clusters: (a) Cachuma, (b) Apostle, (c) Sierra de sangra, (d) Kairakoura, (e) Cedarfire, (f) Cocklebiddy, (g) Holuhraun, (h) Peru

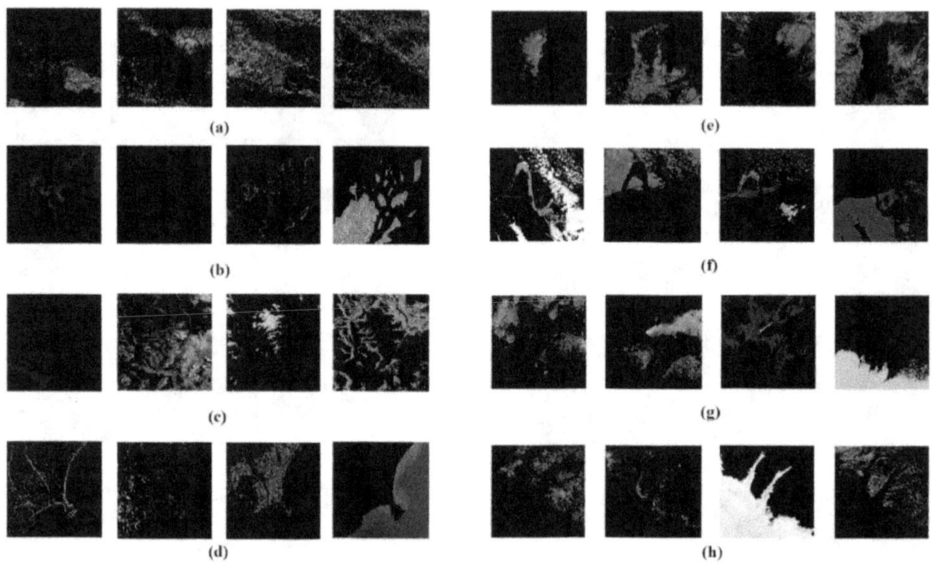

Figure 10. Flowchart of the proposed RDB image segmentation methodology

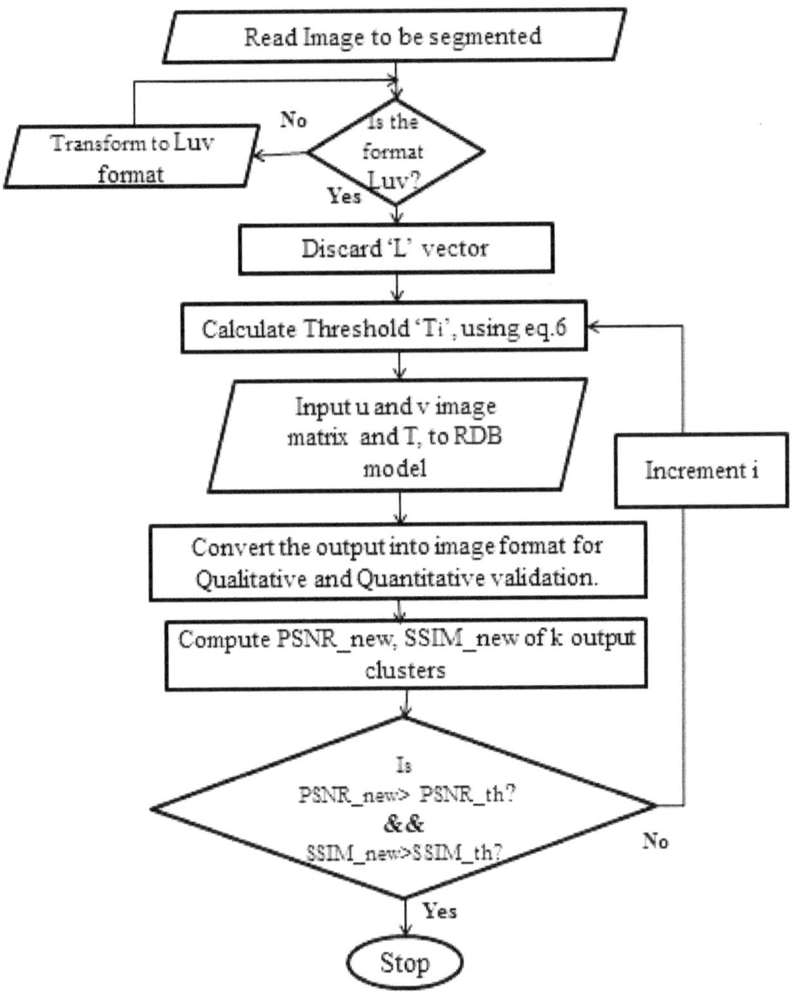

The system simulation of the Behavioral model of the RDBC clustering algorithm has been initially conducted in Matlab 2015a. Convergence of the same is found to be satisfactory, as it is converging after around 50 iterations. Preliminary image segmentation is carried out for Clustering using the standard benchmark Portable Network Graphics (PNG) color image Pepper as shown in the fig.7. Then the testing is further carried out on four Satellite Images of geographical significance (as shown in the fig. 10 and fig. 11), as described below (Demir et al., 2018). The PSNR and SSIM values are plotted (Ratnakumar et al., 2021), comparing the proposed architecture with K-Means and other hardware architectures. Table 1 shows the input satellite images for the simulation processes.

Figure 11. (a) Grouping the satellite image Apostle into four clusters, (b) D matrix plot of apostle_oli islands

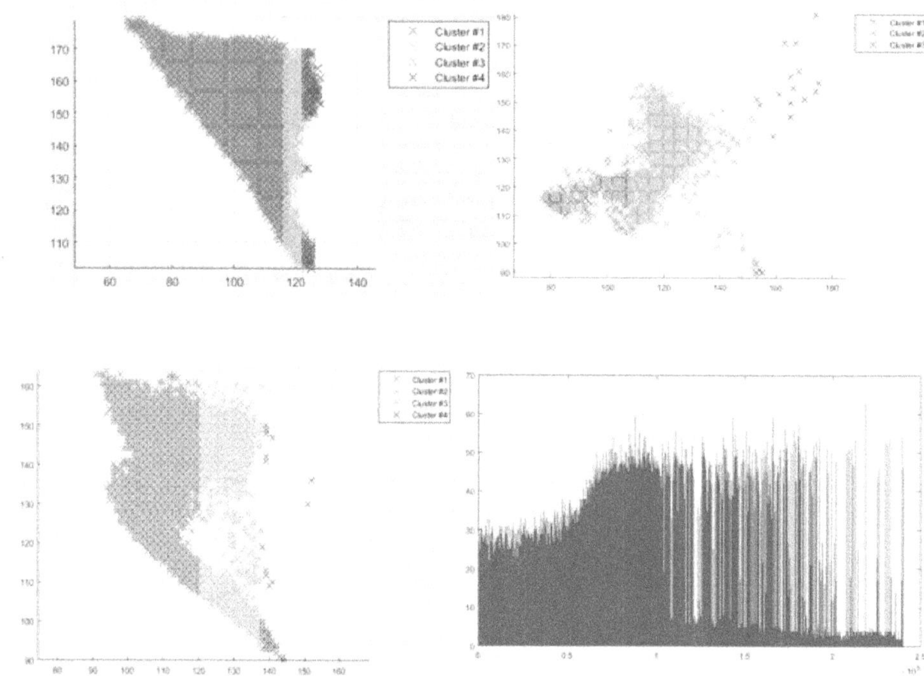

3.1 Discussion on Selected Simulation Cases

Satellite Image Holouhran_oli: This image was captured by the NASA's Operational Land Imager (OLI) on Landsat 8 on September 6, 2014. The ongoing eruptions occurring at Holouhran is clearly depicted in this image (Fig. 8.). The false-color image procured from OLI bands 6-5-3 combines near infrared, shortwave infrared radiations with visible light. The green color represents Ice and the plume of steam and sulphur dioxide appear cyan and bright blue, while liquid water appears navy blue. The Bare or rocky ground around the Holuhraun lava field appears in shades of green or brown in this band combination. The bright orange and red represents fresh lava.

Histogram Analysis of Holouhran_oli: On scrutinizing the histogram of Holouhran_oli it can be understood that there is the possibility of three clusters (Fig. 8).

Cluster Analysis of Holouhran _oli: The image segmentation of Holouhran _oli shows that there can be 3 clusters. The fire and land together constitutes a single cluster, followed by the sea in the second and finally the smoke and river bed in the third and final cluster.

Satellite Image Apostle_oli: It was captured by the NASA's OLI on Landsat 8 on August 14th 2015 of the Islands of Apostle showing 21 islands, at the northern tip of Wisconsin, the place where land touches the Lake Superior. These islands are shown in green color indicating the natural reforestation occurring for many years.

Histogram Analysis of Apostle_oli: The histogram of u and v are dimension of the 'Luv' format. It reveals that there can be predominantly 3 clusters considering the distinct chromatic regions within the image. But for the first two clusters are overlapping (u and v dimensions), the pixel intensity variation is less. So it can be considered to be a single cluster. The initial guess of two clusters can be easily validated, henceforth.

Cluster Analysis of Apostle_oli: These type of satellite images show a proportional increase in the information content as the number of clusters are increased till a limiting value (around 5) is achieved, after that the clusters become undecipherable, reducing the retrieved information content. When the image is clustered into 3 groups, two distinct clusters one containing all the islands and other containing the entire ocean region is formed. The third cluster has no details, which confirms the histogram analysis of the same image.

The overall summary of the comparison gives amazing enlightening on Low-Complexity Hardware for Satellite Image Segmentation for Disaster management and mitigation.

The Challenge is that the Satellite images hold a wealth of information about our planet, but processing them is computationally expensive. K-means clustering, a popular algorithm for grouping similar data points, can be especially demanding.

The Solution lies in the research proposing a custom hardware architecture using a Finite State Machine (FSM) to efficiently implement K-means on satellite images. This "brain" within the hardware guides the processing steps, optimizing resource usage.

Key Features are given below:

Lower Power Consumption: Compared to previous works, this architecture boasts lower power consumption, making it ideal for space applications where energy is at a premium.

Smaller Footprint: It also requires less hardware, using fewer logic units (LUTs), slices, and memory blocks, making it compact and lightweight.

Fast Processing: Despite its simplicity, the architecture maintains a good system clock frequency, ensuring speedy image segmentation.

No Fancy Tricks: Notably, this achievement comes without using any specialized processing blocks, showcasing the efficiency of the design.

Testing Ground: The researchers put their architecture to the test on eight real-world satellite images capturing diverse environmental disturbances, provided by NASA's Operational Land Imager. These images ranged from forest fires to floods, each presenting unique challenges for segmentation.

The Results: The architecture successfully segmented the images, with the quality of the segmentation measured by two key metrics: Peak Signal to Noise Ratio (PSNR) and Mean Structural Similarity Index (M-SSIM).

Beyond the Numbers: What does this mean for real-world applications? By accurately segmenting satellite images, researchers can better understand and respond to environmental disasters. For example, identifying the extent of a wildfire or flood allows for more effective resource allocation and disaster relief efforts.

Looking Ahead: This research paves the way for more efficient and accessible hardware solutions for satellite image analysis. With further development and refinement, this technology could empower us to monitor our planet's health more closely and respond to environmental challenges with greater precision and speed.

In short, this research is a significant step forward in unlocking the secrets hidden within satellite images, allowing us to better understand and protect our planet.

4. CONCLUSION

A review of some of the promising algorithms for performing high speed image segmentation, requiring less computational resources with simple architecture, at the cost of accuracy has been studied. The performance of some algorithms when tested on natural and synthetic images procured from benchmark databases including Landsat Image Gallery of NASA, is found to be excellent considering the parameters like execution time, parallelism and resource utilization. In conclusion, the application of clustering algorithms for image segmentation of satellite images stands as a pivotal advancement in remote sensing and geographical analysis. By employing techniques like k-means, spectral clustering, or hierarchical clustering, these algorithms enable the partitioning of vast image datasets into meaningful and distinct regions based on shared characteristics. This process not only enhances the efficiency of image interpretation but also facilitates the extraction of crucial information from satellite imagery, aiding in land cover classification, urban planning, disaster management, and environmental monitoring. As these clustering algorithms continue to evolve and improve, their integration with satellite image analysis promises to unlock new possibilities for understanding our planet's dynamics and addressing complex global challenges.

REFERENCES

Amit, S. N. K. B., & Aoki, Y. (2017), Disaster detection from aerial imagery with convolutional neural network. *Knowledge Creation and Intelligent Computing (IES-KCIC), 2017 International Electronics Symposium (IEEE),* 239–245 10.1109/KCIC.2017.8228593

Cao, Q. D., & Choe, Y. (2018), Deep Learning Based Damage Detection on Post Hurricane Satellite Imagery. *arXiv preprint arXiv:1807.01688.*

Demir, Koperski, Lindenbaum, Pang, Huang, Basu, Hughes, Tuia, & Raskar. (2018). *Deepglobe: A challenge to parse the earth through satellite images.* ArXiv e-prints.

Doshi, J. (2018). Residual Inception Skip Network for Binary Segmentation. *Proceedings of the IEEE Conference on Computer Vision and Pattern Recognition Workshops*, 216–219. 10.1109/CVPRW.2018.00037

Duarte, D., Nex, F., Kerle, N., & Vosselman, G. (2018). Satellite Image Classification of building damages using airborne and satellite image samples in a deep learning approach. *ISPRS Annals of the Photogrammetry, Remote Sensing and Spatial Information Sciences*, *4*(2), 89–96. doi:10.5194/isprs-annals-IV-2-89-2018

Fujita, A., Sakurada, K., Imaizumi, T., Ito, R., Hikosaka, S., & Nakamura, R. (2017). Damage detection from aerial images via convolutional neural networks. *IAPR International Conference on Machine Vision Applications, Nagoya, Japan*, 8–12. 10.23919/MVA.2017.7986759

He, K., Zhang, X., Ren, S., & Sun, J. (2015). Delving deep into rectifiers: Surpassing human-level performance on imagenet classification. *Proceedings of the IEEE international conference on computer vision*, 1026–1034. 10.1109/ICCV.2015.123

Ignatiev, V., Trekin, A., Lobachev, V., Potapov, G., & Burnaev, E. (2018). Targeted change detection in remote sensing images. *arXiv preprint arXiv:1803.05482.*

Rahman, M. A., & Wang, Y. (2016). Optimizing intersection-over-union in deep neural networks for image segmentation. In *International Symposium on Visual Computing*. Springer. 10.1007/978-3-319-50835-1_22

Ratnakumar, R., & Nanda, S. J. (2019). A low complexity hardware architecture of K-means algorithm for real-time satellite image segmentation. *Multimedia Tools and Applications*, *78*(9), 11949–11981. doi:10.1007/s11042-018-6726-6

Ratnakumar, R., & Nanda, S. J. (2021). A high speed roller dung beetles clustering algorithm and its architecture for real-time image segmentation. *Applied Intelligence*, *51*(7), 4682–4713. doi:10.1007/s10489-020-02067-7

Ratnakumar, R., & Nanda, S. J. (2021, January). An improved genetic clustering architecture for real-time satellite image segmentation. In *2021 International Conference on Advances in Technology, Management & Education (ICATME)* (pp. 123-128). IEEE. 10.1109/ICATME50232.2021.9732768

Ratnakumar, R., & Nanda, S. J. (2023). A high-speed unsupervised hardware architecture for rapid diagnosis of COVID-19. *International Journal of Circuit Theory and Applications*, *51*(1), 437–474. doi:10.1002/cta.3417

Sasaki, Y. (2007). The truth of the F-measure. *Teach Tutor mater, 1*(5), 1–5.

Voigt, S., Giulio-Tonolo, F., Lyons, J., Kucera, J., Jones, B., Schneiderhan, T., Platzeck, G., Kaku, K., Hazarika, M. K., Czaran, L., Li, S., Pedersen, W., James, G. K., Proy, C., Muthike, D. M., Bequignon, J., & Guha-Sapir, D. (2016). Global trends in satellite-based emergency mapping. *Science*, *353*(6296), 247–252. doi:10.1126/science.aad8728 PMID:27418503

Chapter 3
Blockchain Systems in Aerospace Engineering:
Review and Research Challenges

Brindha Subburaj
Vellore Institute of Technology, Chennai, India

Siya Bansal
Vellore Institute of Technology, Chennai, India

Vedansh Kumar
Vellore Institute of Technology, Chennai, India

ABSTRACT

Blockchain is a special computer system that keeps safe and unchangeable records of transactions and ensures records are real, secure, and many people can access them. The way blockchain works is people share their transactions and others make sure they are real. These transactions become a block which is made secure, users agree, and copies are shared. It is widely used for money transfer, internet of things, etc. Recently, it has been applied in the aerospace engineering sector to store and securely manage aircraft records, using blockchain for better communication in aerospace, to perform predictive analytics in battlefield operations to improve the decision making. Research challenges that are to be addressed while extending to aerospace applications include ensuring security and privacy of information and smart contracts, managing deployment cost, handling huge data traffic generated, and rules and regulations specific to this aerospace sector. In this chapter, blockchain concepts, application to aerospace engineering, recent trends, and future research opportunities are discussed.

DOI: 10.4018/979-8-3693-1491-3.ch003

1. INTRODUCTION TO BLOCKCHAIN

Blockchain are open, shared, distributed ledger that helps to record transactions and to track assets between parties of a business network in a verifiable and immutable way. The key feature of Blockchain is that it is decentralised meaning the ledger is maintained by a network of computers all over the world making it more secure and resistant to manipulation. The transaction is verified by network of computers called nodes, before it is entered into ledger, and once it is added it cannot be changed or removed making blockchain immutable for all transaction that have occurred.

Realm of cryptocurrency is most well-known applications in Blockchain. Bitcoin being the most famous and the first application of it. The transaction that happens at this platform are secure and transparent and is added to the blockchain. But it's potential is far beyond cryptocurrency. It can be used to store and verify transaction from the sale of house to the transfer to intellectual property. It also expands to smart contracts, which automatically applies the agreement between parties as it is self-executing.

It has been used in various applications like financial, business and industrial, education, governance, healthcare management, Internet of Things, security and privacy, bigdata management, cloud and edge computing.

1.1 Components of a Blockchain

Blockchain is made up by combining several components such as,

1. Peer-to-Peer (P2P) network: Blockchain are built in the form of P2P network, where direct communication exists between the entities of the network without any central authority. Bitcoin, Ethereum are few examples.
2. Blocks: These are the building blocks of a blockchain. Components of each block is a list of transactions and a different identifier for each transaction which are called a hash. It is generated by cryptographic algorithm, which ensures the probity and security of the information. The detailed components and its purpose are discussed under next section.
3. Chain: To make data secure the blocks are made in chain structure. Every block has the hash of the last block, creating a chronological order of the transactions. Due to this, blocks are in a linked mechanism make it basically resisting any tempering caused.
4. Transactions: These are the actions or exchanges of value recorded on the blockchain. They can include financial transactions, smart contracts, or any other type of digital interaction. Each transaction is verified and added to a

block by miners. The UTXO transactions are used in bitcoin and Account-based transactions are used in Ethereum.

5. Miners: These blockchain network users are responsible for validating and adding new blocks to the chain. To win the right to contribute a block, miners compete to solve challenging mathematical puzzles known as proof-of-work. The security and decentralisation of the blockchain are guaranteed by this procedure.
6. Cryptography: To protect data and uphold privacy, blockchain relies on cryptographic techniques. In order to safeguard the validity and integrity of the transactions, hashing algorithms, digital signatures, and encryption methods are employed. Public key cryptography is used for securing the blockchain environment.
7. Consensus Mechanism: This is the protocol or algorithm that network participants utilise to reach consensus. It makes sure that the authenticity of transactions and the order in which they are added to the blockchain are agreed upon by all nodes in the network. This consensus algorithm is helpful to come to a conclusion of agreeing a common global state among the miners. Proof-of-Work (PoW) and Proof-of-Stake (PoS) are commonly used consensus algorithms.
8. Distributed ledger: It is a ledger of transactions, contracts and are accessible to all the peers in the network. The transaction details are recorded in the ledger in chronological order of its occurrence. The integrity, security and confidentiality of the stored transaction details are preserved. Blockchain is actually a form of distributed ledger.
9. Smart Contracts: These are contracts that self-execute and have predetermined terms and conditions put into the code. They eliminate the need for middlemen and boost efficiency and transparency by automatically executing when the required criteria are met. These are actually programs that are deployed and stored in to blockchains.
10. Wallet: Similar to any real-world account, a wallet is created to handle the Ethereum account. The keys are the actual entity that are stored in a wallet. Many types of wallet are available like custodial wallet, non-custodial wallet, hot and cold wallet, deterministic and non-deterministic wallet, etc.

1.2 Components of a Block in Blockchain

A block is divided in to a header and body. Header contains hash of the previous block, Merkle hash of the transactions, consensus information. Block body contains the transactions. The components of blockchain are as follows:

1. Version number:

This field specifies the version of the blockchain software that was used to create the block.

2. Timestamp:

The time of the creation of the block is determined by this field in blocks. The number of seconds passed after UNIX epoch is what timestamp typically a timestamp represents and this field holds unique value.

3. Nonce:

This is a random number that is used while creating new blocks in blockchain, this process of creating the block is called mining. It involves finding solution to complex problem in mathematics, and the nonce is used as a variable in the problem to generate a hash that meets certain criteria.

4. Difficulty target:

This field specifies the level of difficulty that must be met in order to successfully mine a new block.

5. Merkle root:

The are the hash included in the blockchain which are generated after all the transactions. It is used to ensure that the transaction data in the block is valid and has not been tampered with.

6. Previous block hash:

As to make chain in Blockchain, the hash of pervious block is stored in this flied of the block. This preserves the blockchain's integrity and establishes the blocks' chronological order. The prior block hash field refers to the blockchain's previous block's hash. This field plays a very important role in the security and integrity of the blockchain. By referencing the hash of the previous block, each block in the chain is linked to the block before it, forming a chain of blocks. This helps establish the chronological order of the blocks and ensures that any attempt to modify the contents of a block would be immediately detected.

7. Block hash:

This is a special identification number that was produced using a cryptographic hash algorithm. It is used to identify the block and contributes to maintaining the blockchain's integrity. The block hash is a distinctive identifier that is generated using a cryptographic hash method. This function takes the contents of the block header and transaction data as input and produces a fixed-size output that is unique to the block. The hash code of the block is used to identify the block and helps ensure the integrity of the blockchain. Any effort to change a block's contents would result in a new block hash, which would be promptly noticed.

8. Transaction data:

The transaction data is the main body of the block and contains a set of validated transactions. Each transaction consists of data to prove its trustworthiness, which involves the sender and destination addresses, the amount of cryptocurrency sent, and a digital signature. The transaction data is kept in a Merkle tree, a data structure that makes it possible to verify the data securely and quickly.

Overall, the components of a block in a blockchain network work together to establish the security and integrity of the network. The block header contains metadata about the block, the previous block hash helps establish the chronological order of the blocks, the transaction data contains validated transactions, and the block hash is a unique identifier that helps us to preserve the integrity of the blockchain.

1.3 Genesis Block in Blockchain

The Genesis Block is the foremost block of blockchain and consists of unique characteristics that distinguishes this block from the other blocks in the chain. This is the only block that does not have a previous block hash because there is no previous block present with respect to it. The Genesis Block is hard coded into the blockchain's protocol as the starting point.

The genesis block is created by the blockchain creator or by the agreement of the network's algorithm.

1.3.1 Features of Genesis Block

The following are some of the crucial features of the Genesis Block,

1.No Previous Block:

The Genesis Block is the foremost block in a blockchain, which means that it has no former block for reference.

2.Fixed Block Reward:

In many blockchain networks, the Genesis Block has a fixed block price. This is because the other blocks generally have dwindling block prices, these are designed to motivate miners to continue solving the algorithm to create new blocks.

3.Unique Block Hash:

It has a unique block hash which identifies it uniquely in blockchain, to ensure that genesis block is identified as first.

1.4 Types of Blockchain

Blockchain technology has evolved significantly since its inception, leading to the development of various types of blockchain networks. These networks differ in their design, governance, and access, making them suitable for different use cases. There are mainly four types of blockchains which are Public blockchain, private blockchain, consortium blockchain and Hybrid blockchain

1.4.1 Public Blockchains

Public or permission-less blockchains and are open-source networks that anyone can participate in. In these networks, no central authority or organisation controls the blockchain, meaning that anyone can join the network, create new blocks, and validate transactions. Examples of public blockchains include Bitcoin and Ethereum.

1.4.2 Private Blockchains

Private or permissioned blockchains and are closed networks that are only accessible to authorised users. In these networks, a central authority or organisation controls the blockchain, and users must have permission to join the network, create new blocks, and validate transactions. Private blockchains are often used by businesses or organisations to facilitate internal operations, such as supply chain management or record-keeping.

1.4.3 Consortium Blockchains

Consortium or semi-private blockchains and are controlled by a group of organisations rather than a single entity. In these networks, the consortium members must have permission to join the network, create new blocks, and validate transactions. Consortium blockchains are often used for collaborative efforts between multiple organisations, such as supply chain management or interbank transactions.

1.4.4 Hybrid Blockchains

Hybrid blockchains are networks that combine elements of both public and private blockchains. These networks allow for greater flexibility in terms of access and participation while still maintaining some level of control and security. Hybrid blockchains can be used for a variety of use cases, such as supply chain management or healthcare record-keeping.

2. REVOLUTIONISING AEROSPACE WITH BLOCKCHAIN TECHNOLOGY

In the world of aerospace, building and operating aircrafts is very important, with block chain it will get more safer, organised and efficient. With this technology this industry would change completely. Operators will be able to track past mistakes and faults easily making more efficient and safe crafts to fly in.

Think of blockchain as a super secure digital notebook that lots of people can use at the same time. It's like a magic book where everyone can write down important information, and once it's written, no one can change or erase it. If Blockchain is incorporated with artificial intelligence it can improve the mechanics of the crafts by analysing the past data and incorporating their best features in one craft.

Blockchain can also help in making sure airplanes are always safe to fly. Imagine if all the checks and fixes that planes need to stay in good shape were written down in this special notebook. This would help engineers and pilots know exactly what's been done to keep the planes working well. It's like having a very organised and reliable record of all the work done on each airplane. Air traffic control, which helps planes fly safely without bumping into each other, could also get better with blockchain.

Following of rules and regulation can also get easier with the help of blockchain also It can keep track of things like how much pollution airplanes make or whether they meet safety standards. This makes it easier for everyone to make sure they're doing things the right way.

Using blockchain in aerospace, of course present some issues. One problem is ensuring that various computer systems are capable of communicating with one another and cooperate. Making sure that using blockchain doesn't consume excessive energy or complicate matters is also crucial. Wasim Ahmad et al. (2021) have presented an exclusive and detailed discussions related to extending block chain techniques for aerospace engineering and defense applications.

Future experts are aiming to improve blockchain technology even further for the aircraft industry. They're working out the best ways to use it to make spacecraft and airplanes even more dependable and safe. The aerospace sector could go to new heights with the aid of this unique digital notebook, enhancing everyone's ability to travel safely and effectively.

2.1 Advantages of Blockchain

Imagine blockchain as a superhero with special powers that can change how businesses work. Just like superheroes have unique skills, blockchain has amazing abilities that could transform different areas.

One big power is transparency. Think of blockchain as a clear and open notebook that anyone can look at any time. This is super helpful when being honest and keeping track of things is really important. You could follow products from where they start to where they end up, knowing exactly what happened along the way. This is super useful in industries where having a clear record is a big deal.

Security is another superpower. Blockchain doesn't have one boss; instead, it's like a team of computers that work together. This makes it super tough for bad people to cheat or mess things up. Every time something is added to blockchain, all the computers check to make sure it's real and true. This makes it almost impossible for anyone to cheat or do bad stuff. That's why it's great for things like money transactions.

But wait, there's more! Blockchain also has a power for privacy. Even though everyone can see what's happening, they don't know who's behind it. It's like using secret codes instead of real names. This helps with private transactions, like a secret way to buy things.

When it comes to being strong, blockchain is like a tough shield. It's spread out across lots of computers, so even if something goes wrong with one, the others keep everything safe. This is really important in areas like money and healthcare where things must keep running smoothly.

And guess what? Blockchain can change how businesses work too. Imagine if you could buy things online directly from the seller without middlemen. This would make things faster and cheaper, like a direct path from one person to another.

So, blockchain is a tech superhero with powers like transparency, security, privacy, strength, and changing how business works. Its powers can make things better in many ways.

2.1.1 Advantages in Aerospace Technology

Blockchain technology brings valuable benefits to the aerospace sector. It improves the visibility of supply chains, enabling the live monitoring of parts and materials, which in turn lowers the chances of incorrect information. The secure recording of maintenance and repair histories assists in meeting regulations and boosting aircraft safety. Streamlined by smart contracts, agreements and transactions become automated, simplifying procedures. Effective data sharing among involved parties fosters better collaboration. The unalterable quality of blockchain bolsters security, deterring fraudulent activities, and supports the creation of safe digital identities for aircraft. Moreover, the decentralized storage of data by blockchain can effectively protect sensitive data.

2.2 Disadvantages of Blockchain

2.2.1 Attack on Blockchain

In a blockchain network, a higher percentage of attack occurs when a single entity or a group has control over more than 50% of the network's computing power. This can allow them to manipulate transactions, rewrite history, and potentially double-spend cryptocurrency. While the decentralized nature of most major blockchains makes such attacks highly difficult and costly, smaller or less secure networks could be vulnerable.

2.2.2 Smart Contract Vulnerabilities

Smart contracts, self-executing scripts on a blockchain, are susceptible to bugs or vulnerabilities in their code. Exploiting these weaknesses could lead to unintended consequences or financial losses. The immutability of the blockchain can make fixing such issues challenging once they are deployed.

2.2.3 Private Key Management

Blockchain users access their assets using private keys, which are long strings of characters akin to a password. If a private key is compromised, access to the associated assets is permanently forfeited. Proper management of private keys is crucial, and any compromise could result in significant financial losses.

2.2.4 Initial Coin Offering (ICO) Scams

The popularity of blockchain and cryptocurrencies has led to the rise of fraudulent initial coin offerings (ICOs). Scammers promise investors new cryptocurrencies that never materialize or exaggerate the potential of a project. Investors can lose substantial amounts of money in such schemes.

2.2.5 Exchange Vulnerabilities

Cryptocurrency exchanges, where users trade digital assets, have become targets for hackers due to the large amounts of valuable assets they hold. High-profile exchange hacks have resulted in substantial losses for users. Security vulnerabilities in exchange platforms can lead to unauthorized access and theft.

2.2.6 Regulatory and Legal Uncertainty

The evolving regulatory landscape surrounding blockchain and cryptocurrencies can lead to uncertainty and challenges in terms of compliance. Illicit activities, money laundering, and tax evasion associated with cryptocurrencies have raised concerns among regulators.

2.2.7 Data Privacy

While blockchain transactions are pseudonymous, meaning that they don't directly reveal personal identities, analysis of transaction patterns could potentially unveil user identities. Efforts to enhance privacy, such as with privacy-focused cryptocurrencies, can sometimes conflict with regulatory requirements.

2.2.8 Interoperability and Standardization

The lack of standardization and interoperability between different blockchain platforms can lead to challenges in data sharing and communication between networks. This can hinder the seamless integration of blockchain into existing systems.

2.2.9 Quantum Computing Threat

While still theoretical, the development of powerful quantum computers could potentially undermine the security of current encryption methods used in blockchain systems. This issue is being studied extensively by the cryptographic community to develop quantum-resistant solutions.

2.2.10 Disadvantages in Aerospace Technology

However, challenges exist. Blockchain's computational requirements could be a hurdle for resource-intensive aerospace operations. Integration with existing systems and achieving interoperability may be complex. Privacy concerns arise due to public transaction transparency, despite user identity pseudonymity. The technology's nascent state might necessitate skilled personnel and resource investments for successful implementation. Regulatory uncertainties could also impact adoption. While blockchain offers immense potential for transforming the aerospace industry, addressing these challenges is crucial to realizing its full benefits.

2.3 Security Aspects of Blockchain Systems

The security factors of blockchain technique are to be reviewed and analysed before using it. Few security threats are listed in this section (Li et al., 2020).

- The consensus algorithm is highly vulnerable to security threats.
- The private key security credential is also subject to be compromised, and the ways to secure the private key is still an active research area.
- Bitcoins may indirectly contribute to ransomware, underground market and money laundering, since the user identity information recorded in bitcoin is not related to the real-world information.
- Possibility of user's transaction privacy information leakage.
- Various vulnerabilities associated with smart contract such as timestamp and transaction ordering dependence are to be considered.

3. DRONES AND SWARM MANAGEMENT

Unmanned Aerial Vehicle also known as Drones are applied in various fields such as agriculture, delivery, surveillance, search and rescue operations and defense sector as well. There are various categories of drones available, each capable to fly varied heights and over varied distances guided through a hardware and software setup.

As a progress in drone technology, swarm of drones are used to attain a specific objective and these swarm are known for their scalability and robustness factors. Blockchain techniques are used in various avenues of drones and swarm of drones management. Few notable research in this context is presented in this section.

Nguyen et al. (2022) have used blockchain technique for an Internet-of-Drones system which is used for search and rescue operation. A swarm of drone searches for human presence, observed through the images captured. Blockchain techniques are used to fulfil the security requirements by using Ethereum and hyperledgers. Promising results are presented by the authors shows the future of applying blockchain and AI techniques to such scenarios to further improve their performance. The need for optimizing the usage of the extended blockchain technique is an open research issue for further investigation.

The blockchain technique can be extended to handle recent issues related to the swarm robotics such as security, distributed decision making, behaviour differentiation and adaptation towards latest business models (Ferrer, 2018). Author has discussed, how blockchain technique is useful in securing the communication between the swarm of drones through public key cryptography. The one other emerging research aspect is distributed decision making by all the robots involved in the swarm. Many recent research contributions addressing this issue try to trade-off some quality metric over other to attain the common goal. Using blockchain technique for this decision-making problem is discussed in this paper. Blockchain technique can contribute towards the distributed decision making through techniques like distributed voting systems, collaborative models among the swarm of robots. Trusted systems using blockchain technique to safe and securely use the swarm-as-a service (Wasim Ahmad et al., 2021). Where, the entire process of registration, drone service allocation, metering service and secure amount transactions are all handled using the blockchain system.

4. BLOCKCHAIN IN AEROSPACE AND DEFENSE

Blockchain could be very useful in Airforce defense and other defense department. There are several advantages of blockchain in defense and military like it has decentralized nature, the blockchain's distributed architecture guarantees its availability. Its durability, security, and immutability make it a viable tool for data storage and suggest that it might have a variety of military applications. Aerospace and defense industry executives were surveyed and they found that blockchain if used in defense and military could help in growth of the industry and increase its efficiency (Hartley and Conway, 2018). The prominent areas where we can apply blockchain in military and defense are secure messages, resilient communication, cyber defense, management and access control and international collaboration and trust building. And in this section, we will majorly focus on the previous mentioned topics.

Technically, in the defense field we should use private blockchain, that will be much more useful than a public blockchain. Because in a public blockchain the access to the blockchain would not be contained within the department and might be available for everyone which will make this blockchain very less secure and it defeats our purpose of secure. As we know a private blockchain has barriers to entry and only the administrator can only allow people to access the chain and change its security settings or protocols.

The functioning of military operations is represented as a network model. Sudhan & Nene (2017) have presented one such Network model called Network Enabled Military Operations (NEMO) and extending blockchain technology to this model suitable to be applicable to handle military operations are discussed. Now we are going to explain about how can we use blockchain in defense areas.

4.1 Cyber Defense and Data Integrity

North Korean hackers stole over 237GB of confidential, sensitive facts that belong to South Korea in 2017 (Audrey et al., 2020). The same year, the European Commission said that 80% of European businesses had a cybersecurity concern, and that there were more than 4000 attacks involving ransomware on a daily basis. The economic impact of cybercrime has also risen 5 times in since 2012. US government faced 60,000 cyber-attacks in the energy sector which made it vulnerable since it is also connected to computing technology. If we store data in one place this will become very risky cause this can lead to "terabyte of death" which means that a large amount of classified data being leaked by hackers. The resilient, distributed nature of blockchain can help with detecting and blocking any penetrative attempt being done by the hackers.

On the battlefield, the orders and information received by the soldiers need to be accurate and valid. If digital communication handled by a single organization is going to become more highly susceptible to breaches that could give rise to the communication having its contents intercepted or influenced by the malicious party or malicious users. If a portion of a network is impacted, the security and reliability of the entire system will be jeopardized, and the network may even come to an end. Blockchain can also be the solution for this problem. Considering that we are able to transmit data horizontally, the battlespace becomes more democratic and a safe environment is established due to the fact that if one node fails, other nodes will not be affected.

We can combine conventional military with digitalization of military with respect to of the weapon system. To give one example, the US Navy makes use of blockchain in order to update the out-of-date Aegis Combat System and reinforce its command-and-control network that connects the system's sensors and military equipment. This will enable the weapons to detect and make them fire under millisecond as soon as they see the target.

Another report by Accenture (Hartley and Conway, 2018), points out that inaccurate, manipulated data can cause a large number of consequences like corrupted business insights and wrong decisions being taken on due to this inaccurate data. Additionally, 86% of aerospace and defense companies in Europe and the United States of America want to implement blockchain-based technologies by 2021, according to this survey. Their supply chains and transactions are going to get more secure as a result. The strategy of communal and distributed decision-making security measures processes from cyberattacks. This is because of the secure and non-alterable nature of blockchain. Blockchains will be more reliable cause the security provided by it will not be based on secret or trust or cryptographic keys. This is all due to its transparent and discrete nature.

4.2 Military Logistics and Supply Chain Management

Military logistics and the supply chains have hundreds of sources for military goods and services. With so many individuals, there will inevitably be conflict amongst them, which results in extra expenses, errors, or failure points. So, in order to solve these issues, we can use blockchain. For military purpose we need higher delivery speed, traceability, safety and reduced costs which can be applied with the help of blockchains. The most important feature which can be useful for blockchain is that it is transparent, secure and efficient. This makes it possible for us to confirm that the suppliers in a chain for the armed forces or aerospace sectors are acting legally and are who they state they are.

We can track the components and materials used in the defense sector with the help of blockchain, as well as track the provenance of the weapon from its creation to delivery and stop fakes. We can readily keep track of any previous transactions mainly because of the blockchain's date-stamped framework. This task can be achieved by embedding the sensor, reporting all the processes taking place and with the help of Internet of Things (IoT) we can get the report of each and everything like where its source of material is, how it was delivered, stored, and whether any repairs were made or upgradation was done on it or not. The defense companies can combine the blockchain with AI or quantum computing in order to boost the data analysis or data exchanges.

A far more reliable supply chain could be created if the blockchain could be incorporated into every stage of the manufacturing process, including design, testing, prototype development, and production. The goods can be verified at each step by defense logistics team to get to know what all processes were done for the production of this component or weapon. But right now, due to lack of information may lead to confusion and miscommunication among the participants which leads to lack of trust between them. Nevertheless, if we implement blockchain in this procedure, each supply chain partner will be using a private key, which will make it possible for us to identify the partner. Sensitive data such as the kind of commodities, origin, and destination may be encrypted and only accessible to or visible to a specific participant. For example, Accenture and Thales created a system which is based on blockchain technology, if the aeronautic and defense sectors receive and equipment which is not working properly, they will be able to track details like, when the equipment was received, where it came from and what all processes it went through can easily be found. There could be one more problem like the components could deliberately made faulty or could be made vulnerable or exploitable so we can use blockchain to verify, store and audit the ownership and history of every part of the component. This can ensure transparency and accountability in the defense industry.

In the battlefields, the ability to track the transparency regarding the journey of a product are of the utmost importance. To be successful in its operations, the army must have faith in the prepositioned materials and capabilities. Therefore, with the aid of blockchain, we are able to improve equipment supply chain transparency and also enhance food safety and treatment for injuries on the battlefield. Blockchain can therefore aid in a real-time inventory of the items, materials, and machinery. Rahayu et al. (2021) proposed a military blockchain model to improve the performance of supply chain management for tracking spare parts efficiently, thus improving the military logistics.

Financial losses and waste of materials or goods are also one of the big issues in defense industry and also during the operations. As a result, it is very important to understand where the cash that we invest is going and how it is being utilized. As an example, during the campaigns in Afghanistan, a malfunctioning air surveillance balloon cost 2.7 billion dollars, and $150 million was squandered on private villas for military personnel. So, we are able to avoid similar problems in the future with the aid of money tracking.

4.3 Resilient Communication and Other Uses

Resilient communication could be very useful in army majorly during operations. Blockchain can provide resilient communication. Defense organizations should be ready for adversary electromagnetic spectrum attacks in these kinds of confrontations, particularly those that related to communication systems like satellites, submerged cables, or tactical datalinks. A malicious party may attempt to modify the data in order to break the chain. To combat this threat, military forces must be able to develop, secure, and transmit data, which could possibly be successfully achieved with the aid of blockchain.

Blockchain can help in secure messaging system. The cryptographic encryption mechanism used by blockchains would enable us to add software-based measures that might save costs, integrity is ensured and enhance field communications. During 2016, US launched a project for "Secure Messaging Platform" and for which tenders were launched, the goal of the project was to achieve safely and securely transfer message using a decentralized protocol. NATO is also interested in this project and was showing that they could also help in implementing blockchain solutions. This technology helps to keep the personal data confidential and a digital identity which is stored in blockchain cannot be falsified. The UN used the blockchain to buy food for Syrian refugees, it could also be implemented on defense to create trackers which could track the soldiers during war times or any operation. These trackers can be distributed among all the battalion and can easily be found by the participants which will allow us for real time tracking of the soldiers. Blockchain also offers high accuracy through smart contracts. This can assist us in allocating funding to established programs and unified infrastructure, strengthening supranational defense projects. 3D printing is also being used to construct components for robots, buildings etc. but it can also be used to build drones, gun equipment, shields and other equipment, so blockchain can help in more secure transfer tool for data because of its distributed or decentralized nature.

4.4 International Collaboration and Trust Building

Blockchain's capacity to establish trust and transparency can foster international collaboration within the defense sector. Shared ledgers facilitate secure information sharing and coordination among allied nations, ensuring interoperability and collective security.

For example, Blockchain can be employed to create an interoperable information-sharing platform among NATO member countries. This platform ensures secure and real-time intelligence sharing, enhancing collective defense capabilities and trust among allied nations.

5. SUPPLY CHAIN MANAGEMENT

Supply chain management is adapted in various sectors including aerospace industry. It involves tracking a huge number of components and their supplier details. The conventional web-based method faces difficulty in achieving traceability and transparency (Rahayu et al., 2019). Blockchainallows the stakeholders to first register about the components on the ledger, which makes the traceability easier. The other related application includes, extending blockchain techniques for the procurement process.

The use of blockchains for supply chain management improves the sourcing and manufacturing, inventory management, distribution and delivery, reverse logistics, etc. (Difrancesco et al., 2023). Sunny et al. (2020) have discussed about how blockchain techniques could be used to improve the transparency and traceability of the components during product development in aerospace domain, which helps to improve the trust factor. Blockchain are used find the next time a component would require a service (Hackius & Petersen, 2017).

The manual processing of any procurement in aerospace sector between buyer and seller involves enormous documentation and rate of error in these records are about 10% (Joseph, 2018). Extending blockchain for the aforementioned scenario fulfils the transparency and the reliability properties through digital signatures and other cryptography and hashing techniques. Wasim Ahmad et al. (2021) have discussed about the opportunity of blockchain technology in relation to aviation industry. It is used to secure the loyalty program of the customers, tracking the baggage in real-time and to perform decision making related to billing, auditing the persons who handle baggage. Case studies related to the blockchain platforms for aviation industry like AERON, OZONE are available for further understanding.

6. DEVELOPMENT PLATFORMS FOR BLOCKCHAIN

Here are some detailed examples for blockchain platforms:

6.1 Ethereum

Ethereum is one of the popular blockchain platforms that has been widely used. The smart contract technique is introduced through this platform, where the contracts are self executing and the agreement terms are directly written into code. Ethereum supports the creation and usage of decentralized applications and allows developers to deploy their own tokens and custom smart contracts.

6.2 Tron

Tron is a blockchain platform focused on building a decentralized internet. It aims to provide a platform for content creators to interact with their audiences without intermediaries.

6.3 Hyperledger Fabric

Hyperledger Fabric is an open-source blockchain platform developed by the Linux Foundation. This platform is designed for enterprise related use cases and provides a modular framework for building private and permissioned blockchain networks.

Some more examples for blockchain platforms include, Binance Smart Chain (BSC), Cardano, Polkadot, Solana, Tezos, Avalanche and Algorand.

Here are some beginner-friendly blockchain platforms to consider:

1. Ethereum Testnets (Ropsten, Rinkeby, Kovan): Ethereum, one of the most popular blockchain platforms, offers test networks (testnets) like Ropsten, Rinkeby, and Kovan. These testnets allow users to interact with Ethereum smart contracts and tokens without using real Ether. They provide a safe environment for learning and testing blockchain applications.
2. Truffle: Truffle is again a development framework under Ethereum
3. Ganache: Ganache platform is a personal blockchain emulator that allows users of this platform for creating a local blockchain network for the purpose of testing and development. It provides a graphical interface and helps beginners experiment with smart contracts and DApps without the need for a public blockchain. Much more platforms like Remix, Blockgeeks, Hyperledger Composer Playground, Corda Settler, CryptoZombies, Algorand Studio and Tezos Developer Playground are available as well.

7. OPTIMIZATION IN BLOCKCHAIN

Optimization process try to find solutions that are feasible with respect to one or more objectives considered for evaluation. Application areas regarding the use of optimization includes but not limited to Engineering design optimization, network optimization, supply chain management optimization, block chain optimization, etc. Several methods can be used to perform global search and optimization like the enumerative, deterministic and stochastic methods (Coello et al., 2007). One representative search and optimization technique include Evolutionary Algorithms (EAs). EAs are inspired from the Darwinians principle "survival of the fittest" and the algorithm imitates the natural evolution process. Again, various classes and categories of EAs are available in literature. Optimization problems are characterized using an objective function, and the optimization methods try to find solutions that best solves the function. The problems are classified as single, multi, many objective optimization problems, etc. based on the number of objective functions. Real world problems are mostly multi-objective in nature. Few classic and other promising

Multi-Objective Evolutionary Algorithms (MOEA) include Non-dominated Sorting Genetic algorithm (NSGA-II) (Deb et al., 2002), Non-dominated sorting moth flame optimization (NS-MFO) algorithm (Savsani & Tahwid, 2017), MultiObjective evolutionary algorithm based on decomposition (MOEA/D) (Zhang & Li, 2007), Fuzzy Adaptive Multi-objective Differential Evolution with Diversity Control (FAMDE-DC) (Subburaj & Suthanthira Amalraj, 2021), Strength Pareto Evolutionary algorithm 2(SPEA2) (Zitler et al., 2001), self-adaptive multi-objective differential evolution-based trajectory optimization algorithm (STO) (Subburaj et al., 2023).

Extending such optimization techniques to optimize and solve the multi-objective optimization problems in the Blockchain technique is a latest research area, and the results help to further use this technique for a wide range of use-cases. Few recent research works are presented here. Alofi et al. (2022) have used MOEA to optimize the usage of energy consumption. Trust is built among the users of blockchain is created using consensus algorithm. There are wide range of consensus algorithm available in literature like Proof-of-Work (PoW), Proof-of-Stake (PoS), etc. Practical Byzantine Fault-Tolerant (PBFT), etc. PoW is a widely used consensus technique and its major drawback is the consumption of energy for computation is huge. The authors have used five different MOEA algorithms to minimize the energy consumption and three other related objectives for the efficient usage of PoW. Optimization algorithms like multi-objective particle swarm optimization and strength pareto evolutionary optimization algorithm are applied to optimize and derive optimal block size (Singh & Vardhan 2020). Block size is based on the transactions count within the block, both larger and smaller sized blocks have their own performance issues. Finding optimal block size is characterized as a multi-objective optimization problem and the above evolutionary algorithms are used to solve the same. Extending block chain technique to Internet of Things based applications may demand storage of enormous number of transactions being generated. Extending cloud based storage for storing such transactions and estimating the number of transactions to be moved to cloud based storage platforms using a variant of multi-objective particle swarm optimization algorithm was found to be successful (Nartey et al., 2022).

8. RESEARCH CHALLENGES

The difficulties in extending blockchain techniques for aerospace engineering and related applications are discussed under this section which could be addressed through exhaustive research.

The first and foremost factor is the delay in executing a transaction. The time taken by public blockchain platforms to process and confirm a transaction is higher when compared to a private blockchain platform. The aerospace application domains such as battlefield operations or swarm of drone network, would require decisions to be taken within a stipulated duration making the operation to succeed. Thus, extending block chain techniques to such scenarios ensures seamless operations.

Next, the security and privacy factors are to be considered. Guaranteeing privacy of information is not feasible in public blockchain environments, since all the information are accessible through the ledger. Thus, for mission critical applications like the battlefield operations or other military applications a private blockchain platform is a recommended environment to ensure preserving the privacy of information.

The advancement of IoT equips monitoring and managing real world applications through hundreds of sensors and other related components and devices. This has made generation of huge volume of data traffic flowing in/out the network. For instance, if we consider a drone-based surveillance application and similar scenario may collect real time data and send them to blockchain platform for further analytics like predictive analytics (Druby, 2019). The performance of such predictive analytics techniques highly rely on the volume and variety of data available to perform the analysis. The complex environment of such highly sensitive application may affect the immediate availability of information for further processing which may affect the performance of analytics algorithm. Several researches are carried out related to this context and one alternate solution to this problem is the suggested usage of sidechain blockchains.

The other major factor that is to be considered while extending block chain technique includes, the deployment cost, especially the usage of smart contracts demands a deeper understanding of the blockchain platform to ensure wise usage of the resource. For instance, in Ethereum based block chain platforms, the usage of services are metered and gas charges are to bared by the end user of the network.

Thus, extending block chain techniques to aerospace engineering and other related domain demands further modifications to the underlying base technique to enhance the performance of the application where it is used. Addressing the above performance factors are few open research challenges which require further attention.

REFERENCES

Alofi, A., Bokhari, M. A., Bahsoon, R., & Hendley, R. (2022). Optimizing the Energy Consumption of Blockchain-Based Systems Using Evolutionary Algorithms: A New Problem Formulation. *IEEE Transactions on Sustainable Computing*, *7*(4), 910–922. doi:10.1109/TSUSC.2022.3160491

Audrey, Q., Robin, V., Gilles, W., & Clementz, G. (2020, September 8). *Blockchain in defence: a breakthrough?* https://finabel.org/blockchain-in-defence-a-breakthrough/#:~:text=The%20advent%20of%20blockchain%20technology,the%20technology%20may%20impede%20them

Coello, C., Lamont, G., & Veldhuizen, D. (2007). *Evolutionary Algorithms for Solving Multi-Objective Problems*. Springer.

Deb, K., Pratap, A., Agarwal, S., & Meyarivan, T. (2002). A fast and elitist multiobjective genetic algorithm: NSGA-II. *IEEE Transactions on Evolutionary Computation*, *6*(2), 182–197. doi:10.1109/4235.996017

Difrancesco, R. M., Meena, P., & Kumar, G. (2023). How blockchain technology improves sustainable supply chain processes: A practical guide. *Operations Management Research : Advancing Practice Through Research*, *16*(2), 620–641. doi:10.1007/s12063-022-00343-y

Druby, A. (2019). *Predictive analytics: What and why it matters?* https://www.sas.com/en_in/insights/analytics/predictive-analytics.html

Ferrer, E. C. (2018). The blockchain: A new framework for robotic swarm systems. In *Springer Proceedings of the Future Technologies Conference*. Springer.

Hackius, N., & Petersen, M. (2017). Blockchain in logistics and supply chain: Trick or treat? In *Proceedings of the Hamburg International Conference of Logistics (HICL),* Hamburg University of Technology, Germany. epubli GmbH.

Hartley, C., & Conway, S. (2018, June 12). *Vast Majority of Aerospace and Defense Companies Plan to Integrate Blockchain by 2021, Accenture Report Finds.* https://newsroom.accenture.com/news/vast-majority-of-aerospace-and-defense-companies-plan-to-integrate-blockchain-by-2021-accenture-report-finds.htm

Joseph, F. (2018). *How blockchain will simplify and transform the supply chain.* Accenture White Paper.

Li, X., Jiang, P., Chen, T., Luo, X., & Wen, Q. (2020). A survey on the security of blockchain systems. *Future Generation Computer Systems*, *107*, 841–853. doi:10.1016/j.future.2017.08.020

Nartey, C., Tchao, E. T., Gadze, J. D., Yeboah-Akowuah, B., Nunoo-Mensah, H., Welte, D., & Sikora, A. (2022). Blockchain-IoT peer device storage optimization using an advanced time-variant multi-objective particle swarm optimization algorithm. *EURASIP Journal on Wireless Communications and Networking*, *2022*(1), 1–27. doi:10.1186/s13638-021-02074-3

Nguyen, T., Katila, R., & Gia, T. N. (2023). An advanced Internet-of-Drones System with Blockchain for improving quality of service of Search and Rescue: A feasibility study. *Future Generation Computer Systems*, *140*, 36–52. doi:10.1016/j.future.2022.10.002

Rahayu, S. B., Jusoh, N., Halip, M. H. M., Taib, S. M., & Lee, M. G. (2021). A Conceptual Model of Military Blockchain for Repair Parts Supply Chain Management. In *International Conference on Computer & Information Sciences (ICCOINS)*. IEEE. 10.1109/ICCOINS49721.2021.9497227

Rahayu, S. B., Kamarudin, N. D., & Azahari, A. M. (2019). Military blockchain for supply chain management. *Journal of Education and Social Sciences*, *13*(1), 353–361.

Savsani, V., & Tahwid, A. (2017). Non-dominated sorting moth flame optimization (NS-MFO) for multi-objective problems. *Engineering Applications of Artificial Intelligence*, *63*, 20–32. doi:10.1016/j.engappai.2017.04.018

Singh, N., & Vardhan, M. (2020). Computing Optimal Block Size for Blockchain based Applications with Contradictory Objectives. *Procedia Computer Science*, *171*, 1389–1398. doi:10.1016/j.procs.2020.04.149

Subburaj, B., Jayachandran, U. M., Arumugham, V., & Suthanthira Amalraj, M. J. A. (2023). A Self-Adaptive Trajectory Optimization Algorithm Using Fuzzy Logic for Mobile Edge Computing System Assisted by Unmanned Aerial Vehicle. *Drones (Basel)*, *7*(4), 1–19. doi:10.3390/drones7040266

Subburaj, B., & Suthanthira Amalraj, M. J. A. (2021). A robust and adaptive fuzzy logic based differential evolution algorithm using population diversity tuning for multi-objective optimization. *Engineering Applications of Artificial Intelligence*, *102*, 104240. doi:10.1016/j.engappai.2021.104240

Sudhan, A., & Nene, M. J. (2017). Employability of Blockchain Technology in Defence Applications. In *Proceedings of the International Conference on Intelligent Sustainable Systems (ICISS 2017)*. IEEE. 10.1109/ISS1.2017.8389247

Sunny, J., Undralla, N., & Pillai, V. M. (2020). Supply chain transparency through blockchain-based traceability: An overview with demonstration. *Computers & Industrial Engineering*, *150*, 106895. doi:10.1016/j.cie.2020.106895

Wasim Ahmad, R., Hasan, H., Yaqoob, I., Salah, K., Jayaraman, R., & Omar, M. (2021). Blockchain for aerospace and defense: Opportunities and open research challenges. *Computers & Industrial Engineering*, *151*, 1–14. doi:10.1016/j.cie.2020.106982

Zhang, Q., & Li, H. (2007). MOEA/D: A multiobjective evolutionary algorithm based on decomposition. *IEEE Transactions on Evolutionary Computation*, *11*(6), 712–731. doi:10.1109/TEVC.2007.892759

Chapter 4
Design of Sliding Mode Control Law for Quadrotor With Adaptive Super Twisting Algorithm

Biswapratim Roy
https://orcid.org/0009-0002-1138-1657
National Institute of Technology, Durgapur, India

Aritro Dey
https://orcid.org/0000-0002-2778-5747
National Institute of Technology, Durgapur, India

Jayati Dey
https://orcid.org/0009-0002-8451-7439
National Institute of Technology, Durgapur, India

ABSTRACT

A nonlinear control law is designed for quadrotors that considers actuator saturation, external disturbance, and unknown parameter variations that affect the stability of quadrotors. The proposed robust control law is based on sliding mode control with adaptive super twisting algorithm (SMC-ASTA) and has the sliding surface based on proportional derivative (PD) structure of error while the super twisting algorithm inherits proportional integral (PI) properties. Additionally, the discontinuous control gain of SMC-ASTA law is tuned adaptively as functions of sliding surfaces along with an alternative approach of fuzzy logic tuning. The proposed SMC-ASTA

DOI: 10.4018/979-8-3693-1491-3.ch004

demonstrates improved performance in contexts of dynamic stability, command tracking, and control effort compared to the conventional sliding mode control (CSMC). Asymptotic stability of the proposed SMC-ASTA is theoretically justified with Lyapunov-based stability analysis. Comparative analysis of SMC-ASTA in simulation with the conventional SMC demonstrates its superiority and suggests its suitability for practical applications.

1. INTRODUCTION

In today's active scenario, there is a growing demand for executing complex aerial missions that surpass human capabilities. As a result, Unmanned Aerial Vehicles (UAVs) have risen to prominence, particularly the versatile multi-rotor known as the quadrotor. Quadrotors have garnered substantial attention, proving indispensable in a wide array of sectors spanning defense and civilian applications. Their utility extends across diverse domains, including intelligence operations, surveillance, topographical mapping, industrial inspections, agricultural applications and acting as communication relay platforms, among others (Nakamura et al., 2019).

The complicated structural design of under actuated quadrotors presents formidable challenges when it comes to ensuring stability and control. The inherent nonlinearities in the system dynamics require extensive effort in crafting effective flight control laws. Moreover, even theoretically proved designs, which are designed to guarantee good tracking performance and responsiveness to commands, can sometimes encounter setbacks like partial loss of command authority due to actuator saturation. External factors, such as gusts of wind, and internal variables like variations in payload, further degenerate the stability and maneuvering challenges faced by quadrotors. These complexities emphasize the ongoing exploration in the quest to develop robust control algorithms for quadrotors.

There have been many control laws developed for quadrotors for proper operation and tracking of complex paths. One common approach is linearization of quadrotors dynamics that boils down the complicated nonlinear dynamics into a double integral system dynamics and apply commonly used controllers like Proportional-Integral-Derivative (PID), Linear Quadratic Regulator (LQR), and H_∞ control (Jasim & Gu, 2014; Obias et al., 2019; Shehzad et al., 2019). These linear control methods are favored due to their ease of use and straightforward implementation. However, when external disturbances are present, relying solely on linear approximations can lead to significant negative consequences as the locally linearized model no longer remains corrects after perturbation. To address these issues with linear control methods, design of adaptive feedforward control is also employed. However, it is reported that disturbances still have an impact on the system's performance (Chu et al., 2020).

To tackle this challenge, researchers have explored non-linear control strategies to overcome the limitations of linear methods. One of such nonlinear control technique is continuous sliding mode control (CSMC) and adaptive continuous sliding mode control (ACSMC) have been studied as effective non-linear strategies for controlling quadrotors (Lian et al., 2022; Shao et al., 2022; Tripathi et al., 2022). Additionally, combining fuzzy logic with CSMC has demonstrated considerable improvement in performance (Chen et al., 2021; Huang et al., 2019; Jing et al., 2019; Xu & Lu, 2020). However, a significant drawback of the above control law designs is the requirement of large control efforts, and it can be observed a constant presence of steady state error while the states are tracking the commanded reference.

In the last few years, numerous works has been carried out within the area of adaptive super twisting of sliding mode control, showcasing promising performance in control law (Li et al., 2021). Adaptation is achieved by leveraging the gains of the sliding surface, which is seamlessly integrated with ASTA. In (Ghadiri et al., 2021), a sophisticated approach involves adaptations through the application of a gain matrix determined by the Lyapunov function. Furthermore in (El-Sousy, Amin, et al., 2022), the utilization of a Recurrent Neural Network facilitates the adaptation of the ATSA. Taking into consideration the estimated error of the system in (Ullah et al., 2021), the gain of the discontinuous control within STA is adapted by considering it. In a distinct approach (El-Sousy, Alattas, et al., 2022), the sliding mode control, in conjunction with ASTA, estimates external disturbances and dynamically reconfigures the control effort. This dynamic reconfiguration is accomplished by incorporating a gain matrix that is intricately linked to the estimated disturbance. As the adaptive logic advances, it often leads to increased complexity. As adaptive logic progresses, it frequently results in heightened complexity.

To overcome these obstacles for the maneuvering of the quadrotor while tracking any path with its proper nonlinear dynamics in presences of actuator saturation, external disturbance, and varying model parameters (e.g., payload variation), this chapter presents a new nonlinear control law Sliding Mode Control – Adaptive Super Twisting Algorithm SMC-ASTA with better performance compared to conventional sliding mode control. The proposed SMC-ASTA presented in this work has a few significant features that ensures satisfactory tracking performance at a relatively reduced complexity, implementation rigors. A noteworthy characteristic of the proposed control law lies in the adjustable nature of the discontinuous control gain within the super-twisting algorithm (STA), determined by the sliding surface and its derivative. Here Super Twisting Algorithm (Oliva-Palomo et al., 2019; Seeber et al., 2018)is employed which has Proportional-Integral PI control properties (Anuar et al., 2022; Modi et al., 2023; Zhai & Li, 2022),which eliminates steady state error. This distinctive feature, not previously documented in existing literature, sets the approach apart.

The contributions of the chapter are summarized as follows:

- The proposed non-linear control law SMC-ASTA inherent the properties of PD and PI control logic both, as the sliding surface is constructed based on PD logic and the Super Twisting algorithm is equivalent to PI control.
- In the first case, the discontinuous control gain used in Super Twisting algorithm dynamically adjusts its gains in response to changes in the sliding manifold. In this context, the discontinuous gain of the proposed SMC-ASTA is made adaptive by transforming them into function of the sliding manifolds controlling the quadrotor's four states during its operation.
- In the second case, the discontinuous control gain used in Super Twisting algorithm is decided based on the fuzzy logic having input membership function associated with sliding surface and its derivative.
- Comparative analysis of the altitude and attitude tracking performance of the proposed SMC-ASTA law with the conventional SMC in terms of trajectory tracking, overshoot, response speed, control effort and performance indices.

By combining these elements, the chapter presents a significant advancement in the field of aerospace and flight control systems, offering enhanced adaptability and performance in real-world scenarios.

2. DYNAMICS OF QUADROTORS

This section illustrates the non-linear dynamic model of the quadrotor, utilizing Newton Euler's equation of motion. It has four rotors to exhibit six degrees of motion and for that the quadrotor model is also labeled as an underactuated system. The initial frame (E_i) and the body frame (E_b) help to establish the six degrees of freedom, as illustrated in Fig 1. The state space model (1) of the quadrotor has twelve states that depict six degrees of freedom and demonstrating the non-linear dynamics of the model.

Figure 1. Quadrotor dynamic model

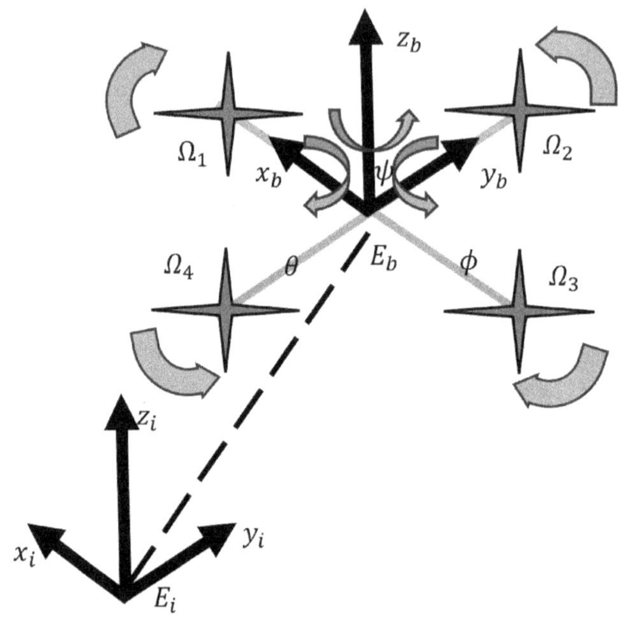

$$x = \left[x, y, z, v_x, v_y, v_z, \phi, \theta, \psi, \omega_\phi, \omega_\theta, \omega_\psi\right] \tag{1}$$

$$\dot{v}_x = \frac{\sin\theta\cos\psi\cos\phi + \sin\psi\sin\phi}{m} U_1 \tag{2}$$

$$\dot{v}_y = \frac{\sin\theta\sin\psi\cos\phi - \cos\psi\sin\phi}{m} U_1 \tag{3}$$

$$\dot{v}_z = \frac{\cos\theta\cos\phi}{m} U_1 - g + D_z \tag{4}$$

$$\dot{\omega}_{\phi} = \omega_{\theta}\omega_{\psi}\left(\frac{I_y - I_z}{I_x}\right) - \frac{Jr}{I_x}\omega_{\theta}\Omega + \frac{l}{I_x}U_2 \tag{5}$$

$$\dot{\omega}_{\theta} = \omega_{\phi}\omega_{\psi}\left(\frac{I_z - I_x}{I_y}\right) + \frac{Jr}{I_y}\omega_{\phi}\Omega + \frac{l}{I_y}U_3 \tag{6}$$

$$\dot{\omega}_{\psi} = \omega_{\theta}\omega_{\phi}\left(\frac{I_x - I_y}{I_z}\right) + \frac{l}{I_z}U_4 \tag{7}$$

The non-linear dynamics of the quadrotor is described by (2)-(7). Where $[\,x, y, z\,]$ are the position in the body frame, corresponding to the position the velocity and the accelerations are $\left[v_x, v_y, v_z\right]$ and $\left[\dot{v}_x, \dot{v}_y, \dot{v}_z\right]$ respectively. The Euler angles are represented by $\left[\phi, \theta, \psi\right]$, mainly roll, pitch and yaw. The angular velocity and the accelerations for the above are represented by $\left[\omega_{\phi}, \omega_{\theta}, \omega_{\psi}\right]$ and $\left[\dot{\omega}_{\phi}, \dot{\omega}_{\theta}, \dot{\omega}_{\psi}\right]$ respectively. The moment of inertias for the positions in the body frame are denoted by $\left[I_x, I_y, I_z\right]$ and the inertia of each rotor is represented by j_r. The mass in the quadrotor is not kept fixed as payload variation is present in the system, so it is denoted by $m = m_{old} + \Delta m$. As external disturbance turbulence is considered which is denoted by D_z. The driving forces for the quadrotor's maneuvering are the thrust produced by the rotors, which are denoted by $\left[U_1, U_2, U_3, U_4\right]$ for altitude climb and attitude changes.

The driving forces for the quadrotor can be formulated as the combination of angular speed of each rotor Ω_{1-4}, the half distance between two arms diagonally denoted by l, the trust factor b and the drag coefficient of the rotors, as illustrated in (8).

$$\begin{bmatrix}\Omega_1^2\\\Omega_2^2\\\Omega_3^2\\\Omega_4^2\end{bmatrix}=\begin{bmatrix}b & b & b & b\\0 & -bl & 0 & bl\\-bl & 0 & bl & 0\\-d & d & -d & d\end{bmatrix}\begin{bmatrix}U_1\\U_2\\U_3\\U_4\end{bmatrix} \tag{8}$$

3. PROBLEM FORMULATION

The state space model of the quadrotor with some external disturbance can be represented as follows (9):

$$\dot{x}=f\left(x\right)+g\left(x,u\right)+w \tag{9}$$

where state vector $\boldsymbol{x}\in\boldsymbol{R}^{12}$, control input, $\boldsymbol{u}\in\boldsymbol{R}^{4}$, the external disturbance $\boldsymbol{w}\in\boldsymbol{R}^{12}$, the nonlinear system dynamics is represented by $f\left(x\right)$ and, $g\left(x,u\right)\in\boldsymbol{R}^{12}$ represents the nonlinear function for the inputs to the quadrotor. Refer Appx.I-Appx.VIII for the final expressions presented:

$$f\left(x\right)=\begin{bmatrix}v_x\\v_y\\v_z\\0\\0\\-g\\\omega_\phi\\\omega_\theta\\\omega_\psi\\r_1\omega_\theta\omega_\psi-r_2\omega_\theta\Omega\\r_3\omega_\phi\omega_\psi+r_4\omega_\phi\Omega\\r_5\omega_\theta\omega_\phi\end{bmatrix} \tag{10}$$

$$g(x,u)=\begin{bmatrix} 0 \\ 0 \\ 0 \\ \left(\cos\phi\sin\theta\cos\psi+\sin\psi\sin\phi\right)U_1/m \\ \left(\cos\phi\sin\theta\sin\psi+\cos\psi\sin\phi\right)U_1/m \\ \left(\cos\theta\cos\phi\, U_1/m\right) \\ 0 \\ 0 \\ 0 \\ q_1U_2 \\ q_2U_3 \\ q_3U_4 \end{bmatrix} \quad (11)$$

$$w=\begin{bmatrix} 0 & 0 & 0 & 0 & 0 & D_Z(t) & 0 & 0 & 0 & 0 & 0 & 0 \end{bmatrix}^T \quad (12)$$

The commanded state trajectory for the quadrotor's altitude is $r_z(t)$ and $\vartheta_z(t)$ as the position and velocity. The commanded Euler's angles and rate of change of Euler's angles are represented by $r_\phi(t), r_\theta(t), r_\psi(t)$ and $v_\phi(t), v_\theta(t), v_\psi(t)$.

$$r(t)=\left[0\,0\,r_z(t)\,0\,0\,\vartheta_z(t)\,r_\phi(t)\,r_\theta(t)\,r_\psi(t)\,\vartheta_\phi(t)\,\vartheta_\theta(t)\,\vartheta_\psi(t)\right]^T \quad (13)$$

The aim of the controller is to track the commanded trajectory $r(t)$. Which requires a robust control law that can asymptotically converge the tracking error. This is mathematically described as (14):

$$\lim_{t\to\infty} x(t)=r(t) \quad (14)$$

4. DESIGN OF CONTROL LAW

An extensive literature review reveals that variants of nonlinear controllers have been designed and employed for better maneuvering of quadrotors since it has highly nonlinear dynamics. Sliding mode control is popular amongst them. However, continuous sliding mode control (CSMC) has some limitations and drawbacks which have been already discussed in the previous section.

Figure 2. Overview of the proposed control law

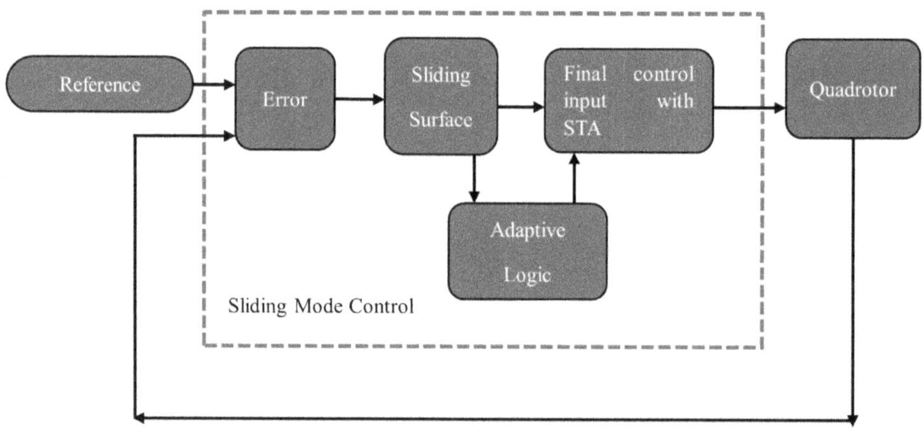

In this section the proposed control law of ASTA with CSMC is presented. The algorithm is developed based on the super twisting framework of sliding mode control wherein the gain of super twisting algorithm is adaptively tuned. The super twisting algorithm requires formulation of a judiciously chosen sliding surface. The adaptation is carried out following two methods, viz., (i) the adaptive logic based of the function of sliding surface and (ii) fuzzy logic tuning. Fig. 2 presents the schematic diagram of the proposed control law which is presented in the subsequent subsections.

4.1 Choice of Sliding Surface

For the proper control structure of CSMC the most vital part is to design the sliding surface of for the CSMC control law. In the present work, since quadrotor is the plant under consideration, four sliding surfaces based on Proportional Derivative (PD) control law are constructed which are denoted by s (15) – (18) for proper control of altitude and attitude with the help of tracking error (16) between the state and the commanded reference.

$$e = r - x \tag{15}$$

$$K = \begin{bmatrix} 0 & 0 & 0 & 0 \\ 0 & 0 & 0 & 0 \\ c_z & 0 & 0 & 0 \\ 0 & 0 & 0 & 0 \\ 0 & 0 & 0 & 0 \\ c_z^{'} & 0 & 0 & 0 \\ 0 & c_\phi & 0 & 0 \\ 0 & 0 & c_\theta & 0 \\ 0 & 0 & 0 & c_\psi \\ 0 & c_\phi^{'} & 0 & 0 \\ 0 & 0 & c_\theta^{'} & 0 \\ 0 & 0 & 0 & c_\psi^{'} \end{bmatrix} \tag{16}$$

$$s = K^T e \tag{17}$$

$$\begin{bmatrix} s_z \\ s_\phi \\ s_\theta \\ s_\psi \end{bmatrix} = \begin{bmatrix} c_z\left(r_z - z\right) \\ c_\phi\left(r_\phi - \phi\right) \\ c_\theta\left(r_\theta - \theta\right) \\ c_\psi\left(r_\psi - \psi\right) \end{bmatrix} + \begin{bmatrix} c_z^{'}\left(\vartheta_z - v_z\right) \\ c_\phi^{'}\left(\vartheta_\phi - \omega_\phi\right) \\ c_\theta^{'}\left(\vartheta_\theta - \omega_\theta\right) \\ c_\psi^{'}\left(\vartheta_\psi - \omega_\psi\right) \end{bmatrix} \tag{18}$$

Here K denotes the gain employed with the construction of sliding surface, where $\left[c_z, \& \ c_z^{'}\right]$ are the gain for the tracking error altitude and $\left[c_\phi, c_\theta, c_\psi, c_\phi^{'}, c_\theta^{'}, c_\psi^{'}\right]$ are the gain for the tracking errors in attitude.

4.2 Structure Control Law

In the proposed sliding mode control law, a modified adaptive Super Twisting Algorithm (ASTA) is incorporated. The advantage of employing STA is that it has the properties of Proportional – Integral (PI) control logic, which can eliminate steady state error in a system. As the sliding surface is based on PD control logic, therefore the proposed control law has both the properties of PD and PI control law. It is illustrated by (19) – (20).

Figure 3. Block diagram of super twisting algorithm

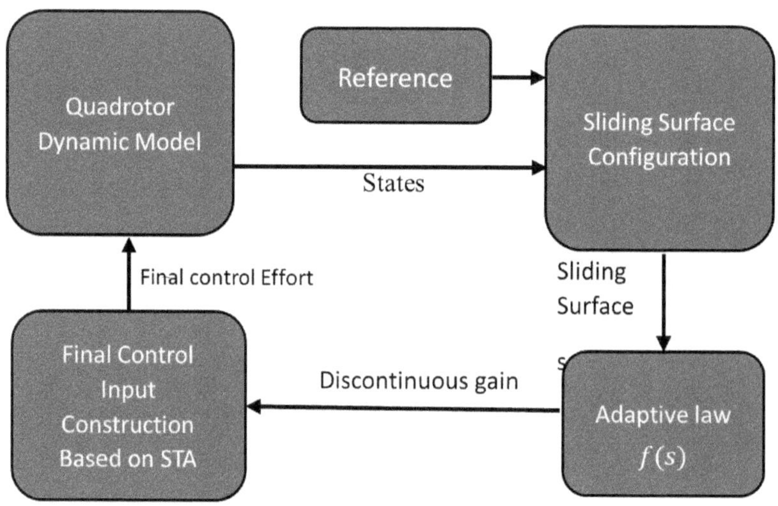

$$\dot{w}_{sta} = -\eta A sign\left(s\right) \tag{19}$$

where $A_{4\times4}$ is a diagonal matrix containing the gains of the discontinuous control for the altitude and attitude of the quadrotor, with a positive constant η. The STA can be described as a two-part control algorithm as it is having $\dot{w}_{sta}$ and U_{sta}. The modification done in the STA is that instead of taking absolute value of the sliding surface and taking the square root value. In the modified STA two-norm of the sliding surface is taken instead as shown in (20).

$$U_{sta} = -\sqrt{A}s_{2} sign\left(s\right) + \int \dot{w}_{sta} \tag{20}$$

4.3 Adaptation of Discontinuous Gain

The discontinuous gain of the proposed control law ASTA is made adaptive using two methods. In the first method the discontinuous gain depends on the function of sliding surface and its derivative. For the second method the discontinuous gain depends on a fuzzy logic-based rule where the input membership functions are based on sliding surface, absolute error, and their derivatives.

4.3.1 Adaptation Based on Function of Sliding Surface

In this method the adaptation of the discontinuous gain is done by a function which is designed with help of sliding surface s_i and its derivatives $\dot{s}_i$. When the state trajectory deviates from the sliding surface, the discontinuous gain changes its magnitude to bring the state trajectory to the sliding surface enabling proper control and reference tracking of the quadrotor. The mathematical expression of the function based on the sliding surface and its derivative can be expressed as (21):

$$\widehat{A} = \left\{ \left(\alpha \left| s \right| + \beta \left| \dot{s} \right| \right) + \gamma \right\} \tag{21}$$

$$\widehat{A} = \begin{bmatrix} a_1 & 0 & 0 & 0 \\ 0 & a_2 & 0 & 0 \\ 0 & 0 & a_3 & 0 \\ 0 & 0 & 0 & a_4 \end{bmatrix} \tag{22}$$

where $\widehat{A}_{4\times4}$ is the new diagonal matrix containing the discontinuous control gains for the altitude and the attitude of the quadrotor. For providing different weightage, fine tuning, and scaling the sliding surface and its derivatives two positive integer constant $\alpha > 0$ and $\beta > 0$ are used. At instance when the state trajectory is laying in the sliding surface the function might set the magnitude of the discontinuous gain to zero, to avoid such problem a position integer $\gamma > 0$ is used that doesn't let the magnitude of the discontinuous gain to ever reach zero.

Figure 4. STA with function of S f(s)

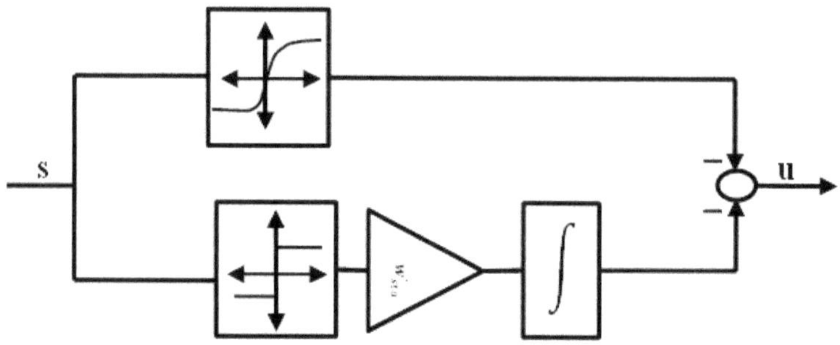

4.3.2 Adaptation Based on Fuzzy Logic

The discontinuous control gain is determined by a function (23) based on fuzzy logic in this section. The function for adaptation of the discontinuous control gain contains two variables that a regulated by fuzzy rules. The input membership function considered for the fuzzy rules are based on sliding surface and absolute error with their derivatives.

$$\widehat{A} = (\sigma \xi + \Gamma) \tag{23}$$

Here $N_{i_{4\times4}}$ is the new diagonal matrix containing the discontinuous control gains for the altitude and the attitude of the quadrotor. The function has two parts where, σ_i is determined by the fuzzy rule in Table 1, based on the membership function of sliding surface and its derivative having three sets negative (NN), Zero (ZZ) and positive (PP), as shown is Fig 6, where Γ is regulated by the fuzzy rule in Table 2. The fuzzy rule is constructed with the membership function of absolute error and its derivatives containing three sets small (SS), medium (MM) and big (BB), as shown is Fig 7. To scale and fine tune σ_i, a very small positive value $\xi > 0$ is used.

Figure 5. STA with fuzzy logic

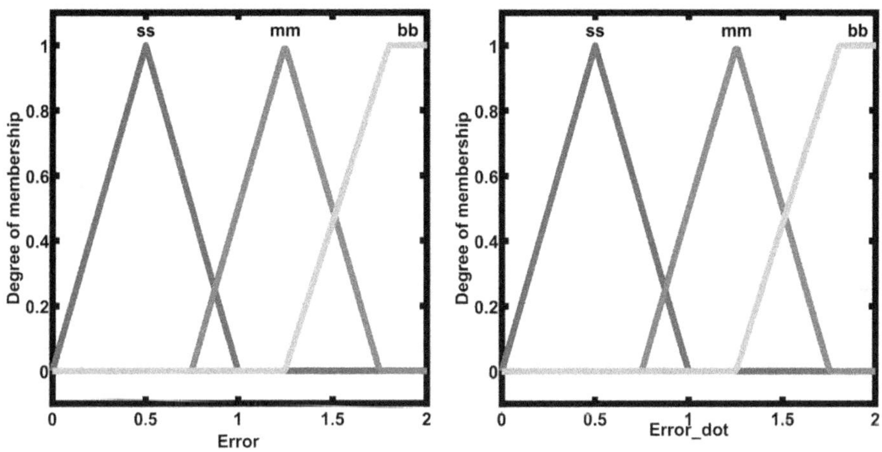

Table 1. Fuzzy rule for σ

$\dot{s}_i$	s_i		
	NN	*ZO*	*PP*
NN	SIG3	SIG1	SIG2
ZO	SIG2	SIG1	SIG2
PP	SIG2	SIG1	SIG3

Figure 6. Fuzzy membership function based on sliding surface

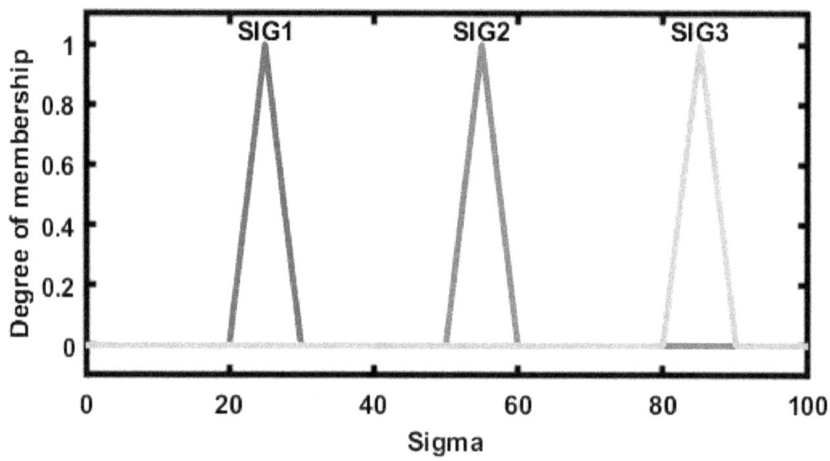

Table 2. fuzzy rule for Γ

e	*e*		
	SS	*MM*	*BB*
SS	GA3	GA1	GA2
MM	GA1	GA2	GA3
BB	GA2	GA3	SIG3

Figure 7. Fuzzy membership function based on error

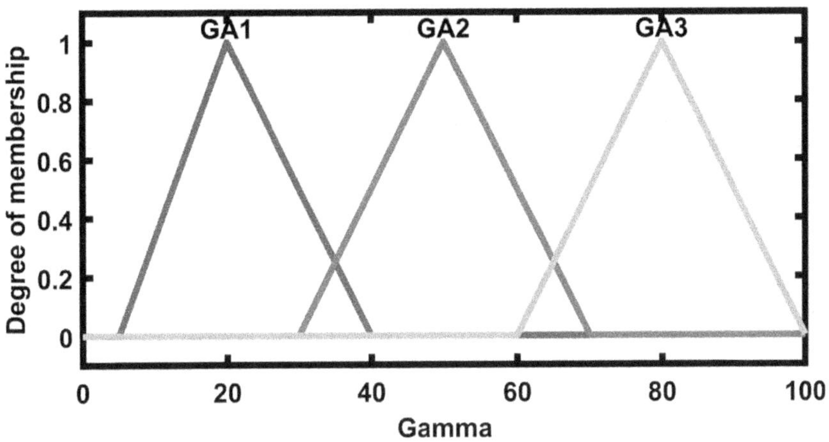

Figure 8. Output Membership of σ

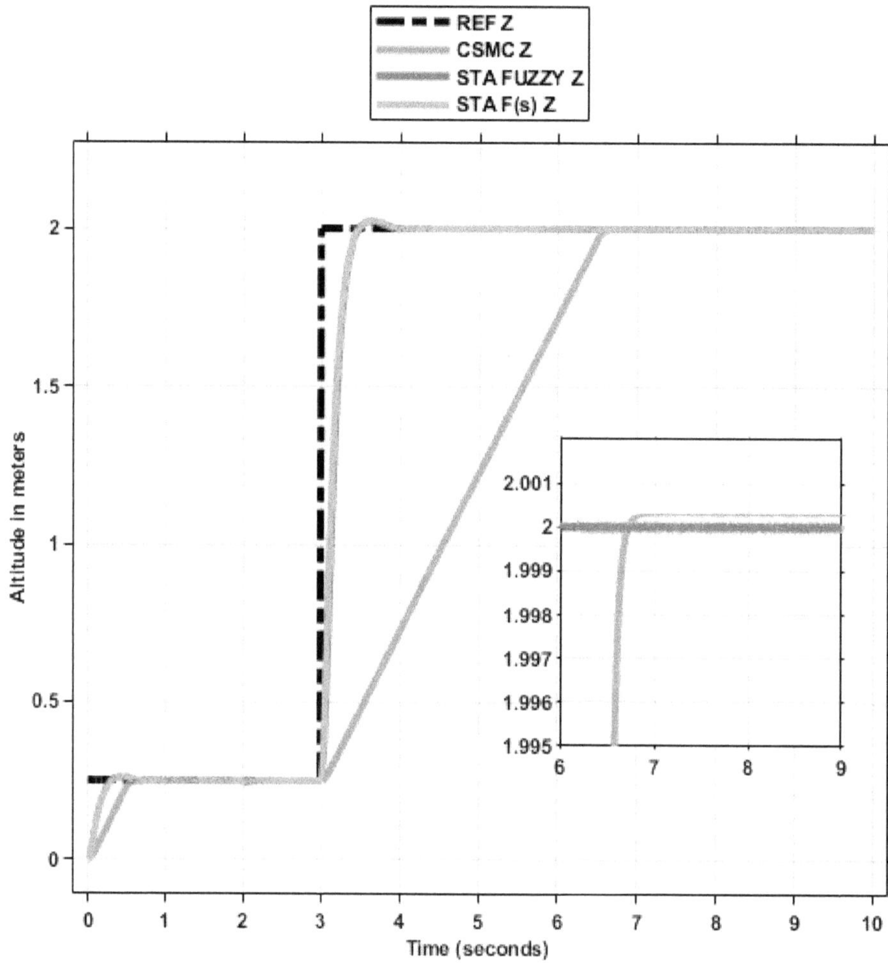

Figure 9. Output Membership of Γ

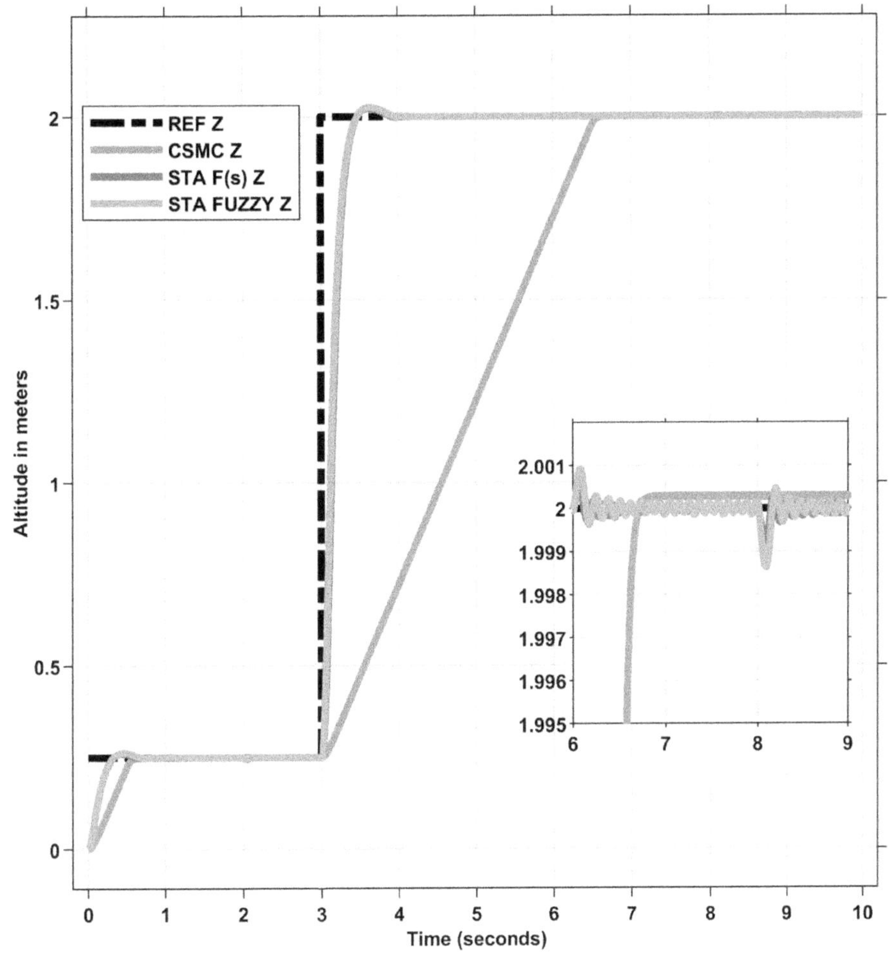

4.4 Final Design of Control Law Based on Super Twisting

For the quadrotor to operate and track the commanded reference for the altitude and attitude, it requires four control inputs. To formulate the desired control inputs, the modified ASTA is incorporated as the reaching law of the proposed control law (27). First the derivative of the sliding surface is calculated (24) and substituted in (27):

$$\dot{s} = K^T e \tag{24}$$

$$\begin{bmatrix} \dot{s}_z \\ \dot{s}_\phi \\ \dot{s}_\theta \\ \dot{s}_\psi \end{bmatrix} = \begin{bmatrix} c_z\left(\dot{r}_z - \dot{z}\right) \\ c_\phi\left(\dot{r}_\phi - \dot{\phi}\right) \\ c_\theta\left(\dot{r}_\theta - \dot{\theta}\right) \\ c_\psi\left(\dot{r}_\psi - \dot{\psi}\right) \end{bmatrix} + \begin{bmatrix} c'_z\left(\dot{\vartheta}_z - \dot{v}_z\right) \\ c'_\phi\left(\dot{\vartheta}_\phi - \dot{\omega}_\theta\right) \\ c'_\theta\left(\dot{\vartheta}_\theta - \dot{\omega}_\theta\right) \\ c'_\psi\left(\dot{\vartheta}_\psi - \dot{\omega}_\psi\right) \end{bmatrix} \tag{25}$$

$$\dot{s} = -\sqrt{\widehat{A}s_2}sign\left(s\right) + \int \dot{w}_{sta} \tag{26}$$

$$\dot{s} = -\sqrt{\widehat{A}s_2}sign\left(s\right) - \int \eta\widehat{A}sign\left(s\right) \tag{27}$$

$$K^T\dot{e} = -\sqrt{\widehat{A}s_2}sign\left(s\right) - \int \eta\widehat{A}sign\left(s\right) \tag{28}$$

$$K^T\left(\dot{r} - f\left(x\right) - g\left(x, u\right)\right) = -\sqrt{\widehat{A}s_2}sign\left(s\right) - \int \eta\widehat{A}sign\left(s\right) \tag{29}$$

$$K^T\dot{r} - K^Tf\left(x\right) - K^Tg\left(x, u\right) = -\sqrt{\widehat{A}s_2}sign\left(s\right) - \int \eta\widehat{A}sign\left(s\right) \tag{30}$$

$$K^Tg\left(x, u\right) = K^T\dot{r} - K^Tf\left(x\right) + \sqrt{\widehat{A}s_2}sign\left(s\right) + \int \eta\widehat{A}sign\left(s\right) \tag{31}$$

Solving the differential equation (31) for the 3rd of the quadrotor the final control input can be expressed as (32):

$$U_1 = \left[c_z^{'} \frac{m}{\cos\theta \cos\phi_1} \right]^{-1} \left[c_z^{'} g + c_z \dot{r}_z - c_z v_z - c_z z + \sqrt{a_1} s_{z2} sign\left(s_z\right) + \int \eta a_1 sign\left(s_z\right) \right] \tag{32}$$

Likewise solving for the 7th, 8th and the 9th state of the quadrotor, one obtains the following final control inputs (33)-(35):

$$U_2 = \left[c_\phi^{'} q_1 \right]^{-1} \left[c_\phi \dot{r}_\phi - c_\phi \phi - c_\phi \omega_\phi + c_\phi^{'} \left(r_2 \omega_\theta \Omega - r_1 \omega_\theta \omega_\psi \right) + \sqrt{a_2} s_{\phi 2} sign\left(s_\phi\right) + \int \eta a_2 sign\left(s_\phi\right) \right] \tag{33}$$

$$U_3 = \left[c_\theta^{'} q_2 \right]^{-1} \left[c_\theta \dot{r}_\theta - c_\theta \theta - c_\theta \omega_\theta - c_\theta^{'} \left(r_3 \omega_\phi \omega_\psi + r_4 \omega_\phi \Omega \right) + \sqrt{a_3} s_{\theta 2} sign\left(s_\theta\right) + \int \eta a_3 sign\left(s_\theta\right) \right] \tag{34}$$

$$U_4 = \left[c_\psi^{'} q_3 \right]^{-1} \left[c_\psi \dot{r}_\psi - c_\psi \psi - c_\psi \omega_\psi + c_\phi^{'} r_5 \omega_\theta \omega_\phi + \sqrt{a_4} s_{\psi 2} sign\left(s_\psi\right) + \int \eta a_4 sign\left(s_\psi\right) \right] \tag{35}$$

4.5 Stability Assessment

The Stability assessment of the proposed control law is verified with the help of Lyapunov second principle on asymptotic stability analysis. Disturbance w and modelling uncertainties are considered in the state equation of quadrotor (13). The Lyapunov function V(t) is necessarily required to be positive definite and its first derivative must be negative definite to ensure asymptotic stability of the proposed control law:

$$V = \frac{1}{2} s^T s \tag{36}$$

$$\dot{V} = \frac{1}{2} s^T \dot{s} + \frac{1}{2} \dot{s}^T s \tag{37}$$

It is to be noted that during construction of the final control input to the quadrotor the external disturbance is ignored. With the consideration of the disturbance (13). The reaching law (27) can be written as (40):

$$\dot{s} = K^T e - K^T w = -\sqrt{\widehat{A} s_2} sign\left(s\right) - \int \eta \widehat{A} sign\left(s\right) \tag{38}$$

$$\dot{s} = -\sqrt{\widehat{A} s_2} sign\left(s_i\right) - \int \eta \widehat{A} sign\left(s\right) + K^T w \tag{39}$$

where $K^T w$ is a vector function of bounded external disturbance which is considered to be a step input in this work. For that reason, the rate of change of the disturbance $K^T w$ is within very small magnitude.

$$\begin{aligned}\dot{V} &= \frac{1}{2} s_i^T \left(-\sqrt{\widehat{A} s_2} sign\left(s_i\right) - \int \eta \widehat{A} sign\left(s\right) + K^T w\right) \\ &+ \frac{1}{2}\left(-\sqrt{\widehat{A} s^T{}_2} sign\left(s^T\right) - \int \eta \widehat{A} sign\left(s^T\right) + w^T K\right) s_i\end{aligned} \tag{40}$$

$$\dot{V} = \frac{1}{2}\left(-\sqrt{\widehat{A} s_2} \sum_{i=1}^{n} \left|s\right| - \eta \widehat{A} t \sum_{i=1}^{n} \left|s\right| + s^T K^T w\right) + \frac{1}{2}\left(-\sqrt{\widehat{A} s_2} \sum_{i=1}^{n} \left|s\right| - \eta \widehat{A} t \sum_{i=1}^{n} \left|s\right| + w^T K s\right) \tag{41}$$

$$Q = \left(s^T K^T w + w^T K s\right) \tag{42}$$

$$\dot{V} = \left(-\sqrt{\widehat{A} s_2} \sum_{i=1}^{n} \left|s\right| - \eta \widehat{A} t \sum_{i=1}^{n} \left|s\right| + \frac{1}{2} Q\right) \tag{43}$$

The magnitude of discontinuous gains should be chosen such that it exceeds the value of the external disturbance. Hence, it is clearly observed from (44) that with proper adaptation of the discontinuous control gain the Lyapunov function (38) even in presence of external disturbance and ensures asymptotic stability of the closed loop system.

5. SIMULATION RESULTS

The proposed control law ASTA in the book chapter is validated by simulation. The chosen model parameters pertain to a typical mid-sized quadrotor, accommodating a payload fluctuation of approximately $\pm 25\%$. This setup is employed to evaluate the effectiveness of the proposed control law in scenarios where the mass is not constant and is under external disturbance. Given that Brushless DC (BLDC) motors are the customary choice for generating thrust in quadrotors, the BLDC motors' speed range under consideration spans approximately ± 1000 radians per second.

Table 3. Parameter for BLDC motor

Parameters	Value
Mass (m)	2kg
Gravitational force (g)	9.81 m/s^2
Turbulence (D_z)	50% variation in the reference altitude
Momemnt of Inertia of the rotor (J_r)	$2*10^{-5}$ $kg.m^2$
Momemnt of Inertia about x axis (I_X)	0.0259 $kg.m^2$
Momemnt of Inertia about y axis (I_Y)	0.0260 $kg.m^2$
Momemnt of Inertia about z axis (I_Z)	0.0397 $kg.m^2$
The arm length (l)	0.285 m
The thrust factor (b)	$1.264*10^{-4}$ $kg.m^2$
The drag factor (d)	$6.5*10^{-5}$ $Nm.s^2/rad^2$

The propose control law is subjected to, two different scenarios first the commanded reference are chosen as step signals and in the second scenario the quadrotor is subjected to a helical path to track the trajectory properly. A comprehensive comparative analysis is conducted involving established sliding mode and fuzzy sliding mode laws, in conjunction with the proposed ASTA law. The helical path is described by (44)-(46) $z_d = r_z(t)$ and $r_\psi(t)$ kept constant.

$$x_d = 10\cos 0.1\text{t} + 10 \tag{44}$$

$$y_d = 10\sin 0.1\text{t} + 10 \tag{45}$$

$$z_d = 0.1\text{t} \tag{46}$$

5.1 Validation With Step Command

The altitude responses for the conventional sliding mode control and the proposed control law ASTA with two separate ways to regulate the discontinuous control gain is shown in Fig 10. and with presence of external disturbance, it is shown in Fig 11. The roll, pitch, and the yaw angle tracking of the step commanded reference is shown in Fig 12 – 14. The control effort required for the conventional sliding mode control and the proposed control law ASTA is illustrated in Fig 15.

Figure 10. Altitude tracking for step input

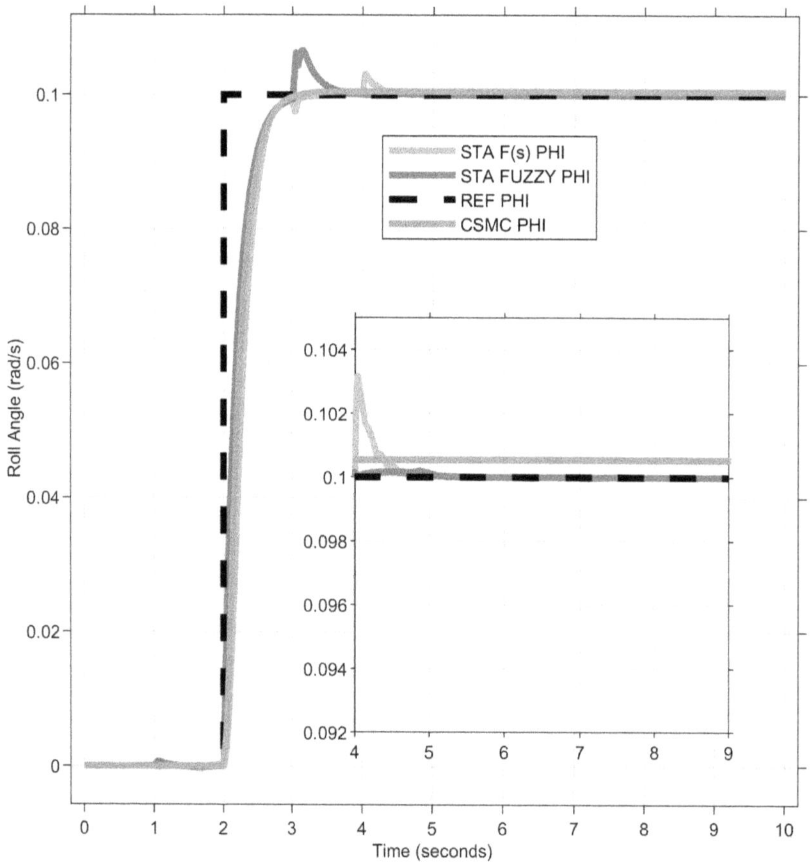

Figure 11. Altitude tracking for step input in presence of disturbance

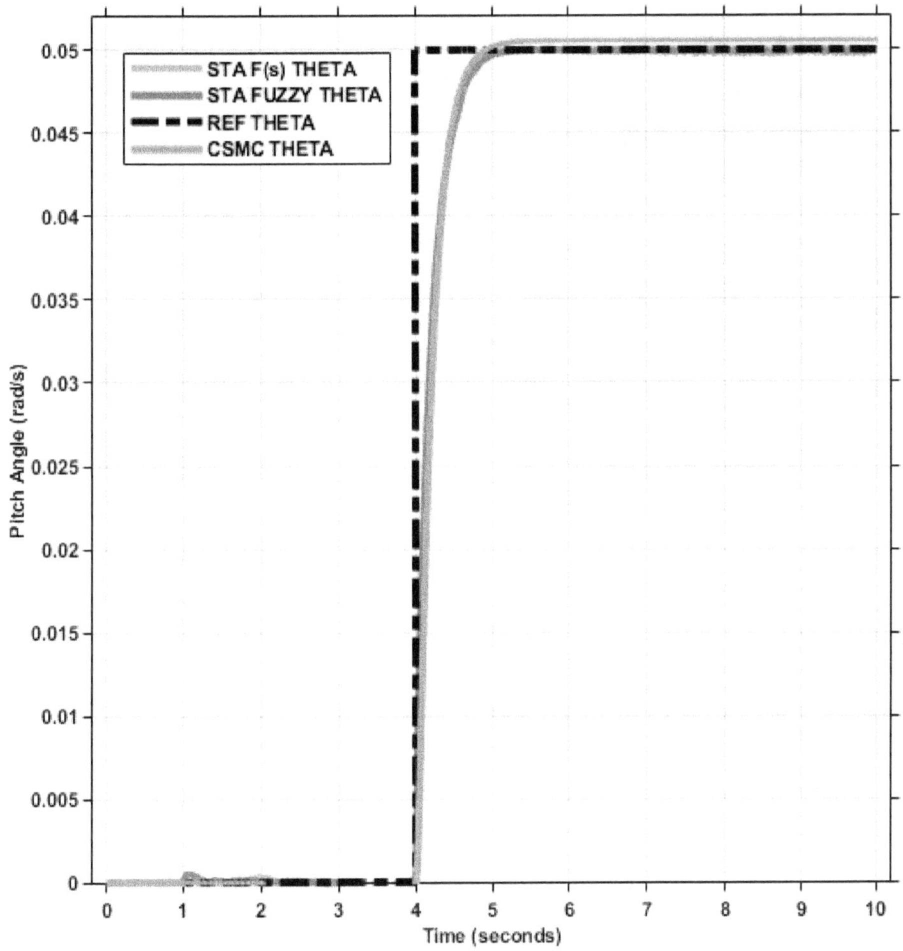

Figure 12. Roll angle tracking for step input

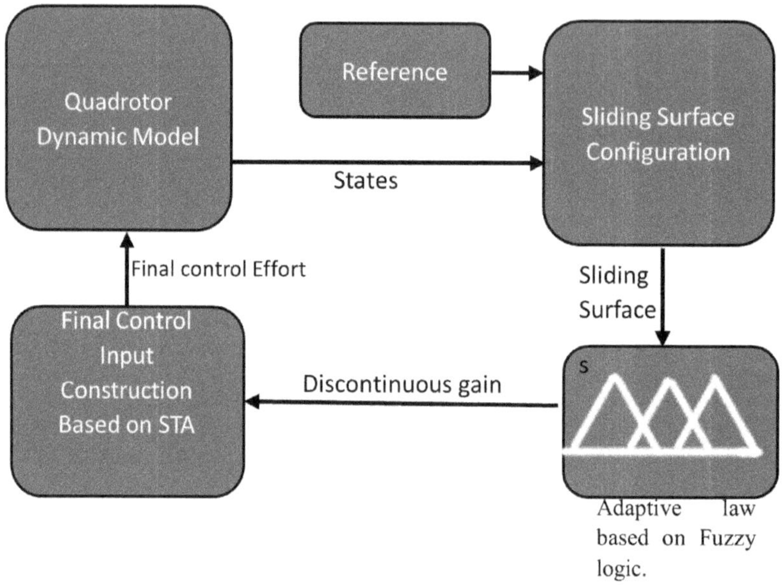

Figure 13. Pitch angle tracking for step input

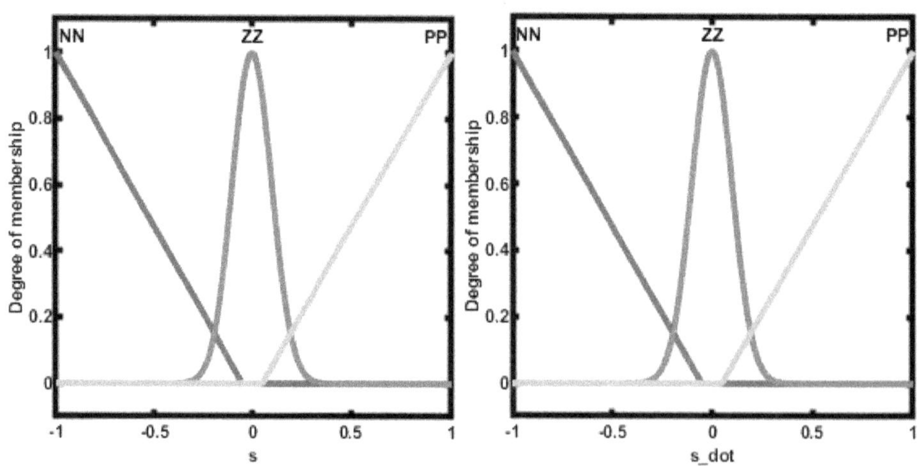

Figure 14. Yaw angle tracking for step input

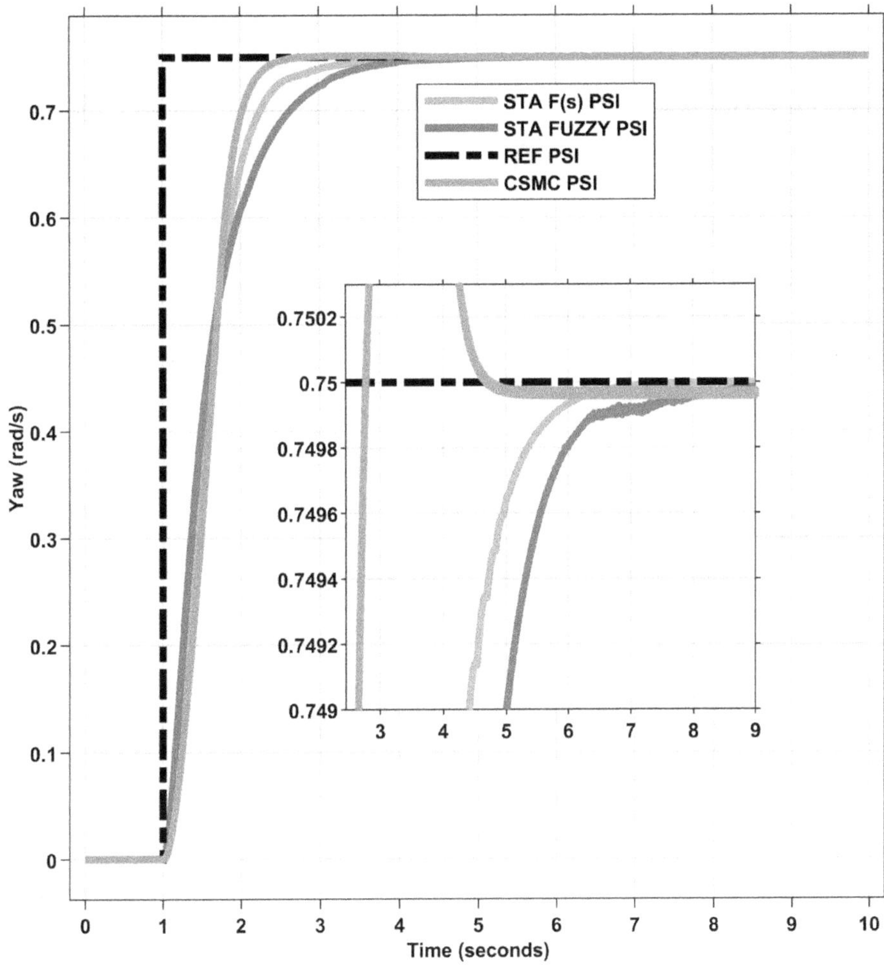

Figure 15. control efforts for step commanded reference

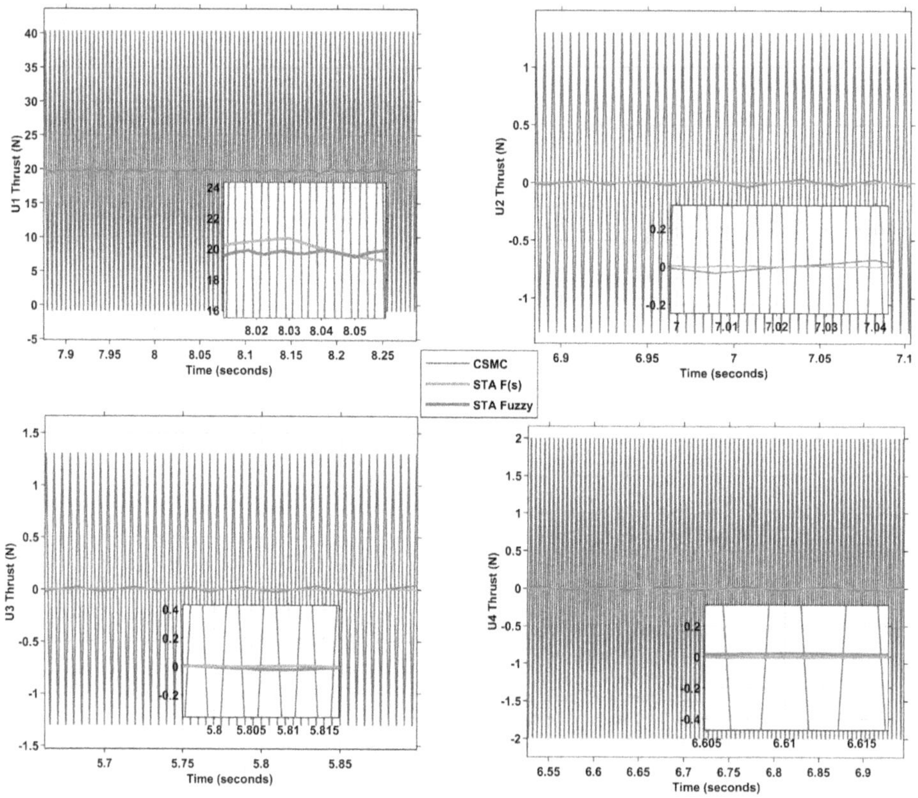

Table 4. Performance index comparison

Performance Index	Control Laws With Disturbance for Step Commanded Reference		
	CSMC	*ASTA f(s)*	*ASTA Fuzzy*
DOC	1.7315	1.4086	**1.4078**
Lu	3.23e+3	**159.59**	172.1957
Lud	5.60e+3	**224.80**	242.4131
ISE	0.6262	**0.0985**	0.0988
IAE	0.5786	**0.1790**	0.1798
ITAE	2.0152	0.3884	**0.3883**

From the above Fig 10 – 114 it is evident that CSMC law is having a prominent steady error. It is also observed that in Fig 10- 11 the settling time is much better than the proposed control laws. From Fig 15 it can be observed that the control efforts are more when CSMC is employed than the proposed control laws with more chattering frequency. In Table 4. the comparison among the control laws by the performance indices proposed by authors in (Yogi et al., 2021), states that the chattering is more in the CSMC than the proposed control laws and in terms of error minimization the proposed control law performed better.

5.2 Trajectory Tracking

The helical path tracking is shown in in Fig 16. And the altitude tracking of the helical path is, it is shown in Fig 17. The roll, pitch, and the yaw angle tracking of the helical path tracking is shown in Fig 18 – 20. The control effort required for the conventional sliding mode control and the proposed control law ASTA is illustrated in Fig 20.

From the above Fig 17 – 20 it can be observed that a constant presence of steady state error is prominent for the CSMC control law and high overshoots are also observed form Fig 18. The tracking performance of the proposed control laws is comparatively better than the CSMC. The reduction in chattering among the control efforts is also observed for the proposed control laws from Fig 21.

Figure 16. Helical path tracking

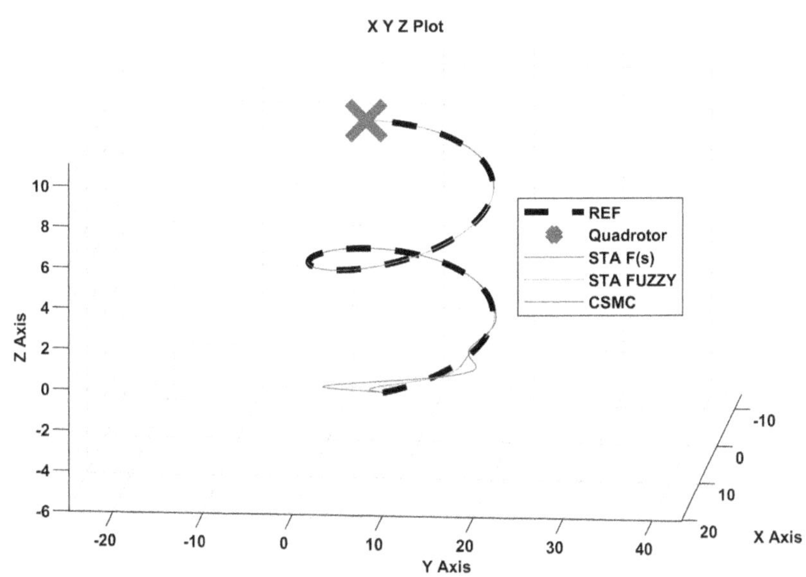

Figure 17. Altitude tracking of the helical path

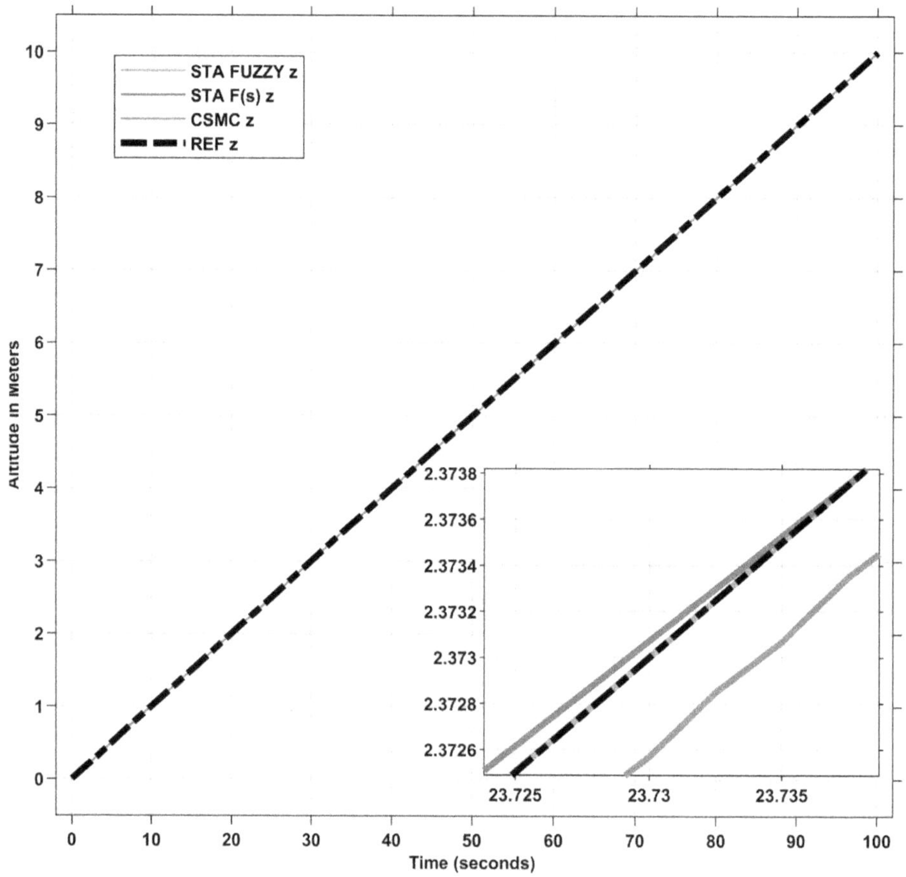

Figure 18. Roll angle tracking of the helical path

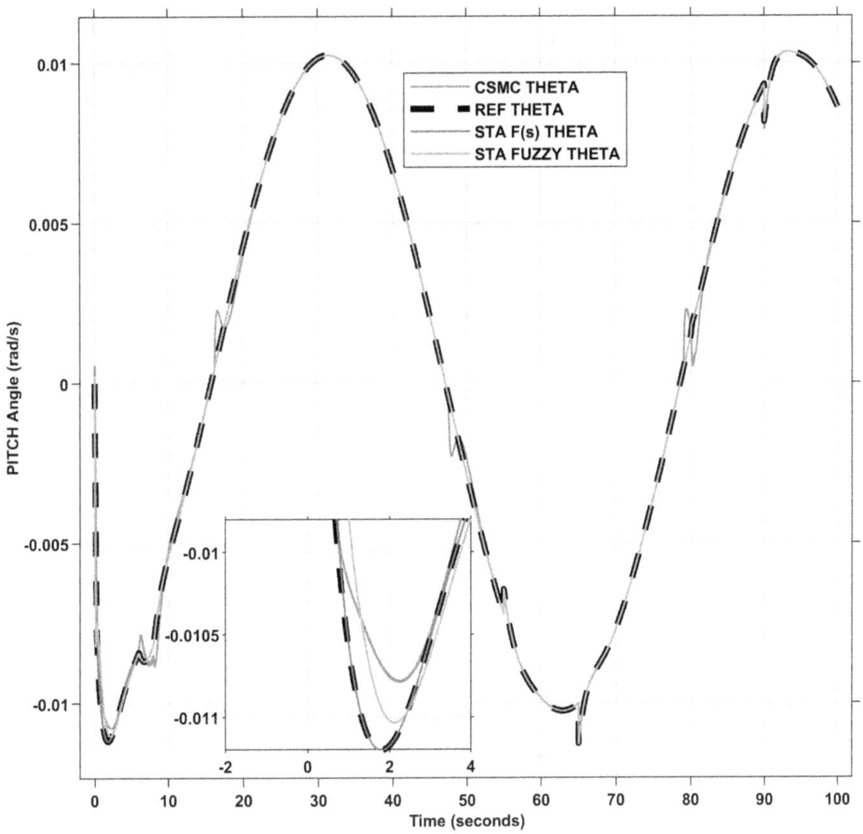

Figure 19. Pitch angle tracking of the helical path

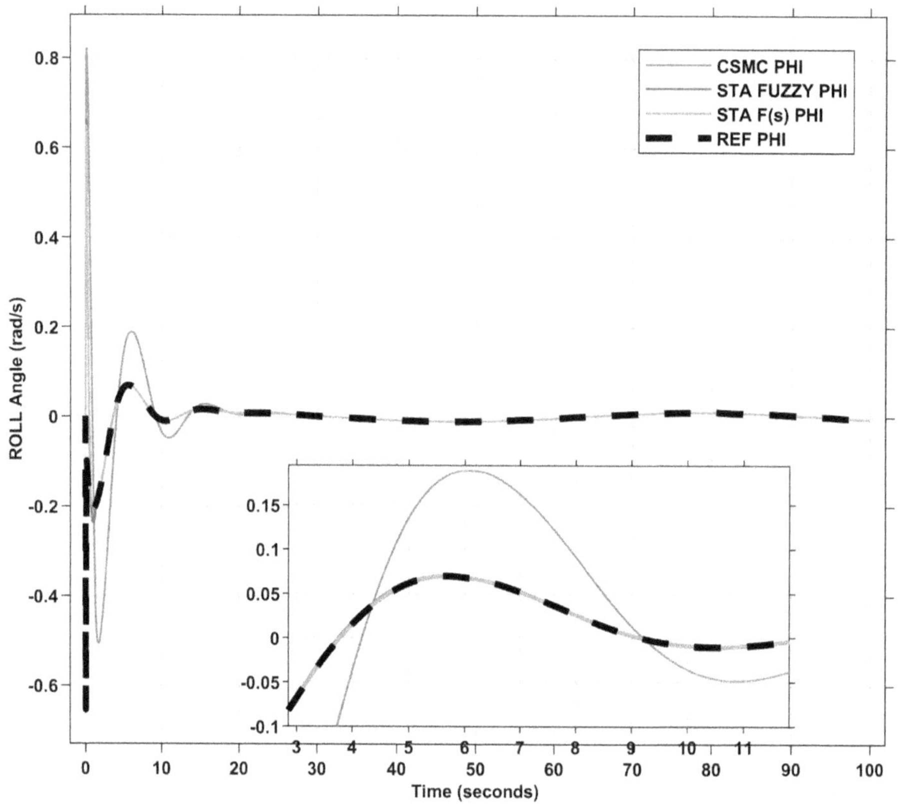

Figure 20. Yaw angle tracking of the helical path

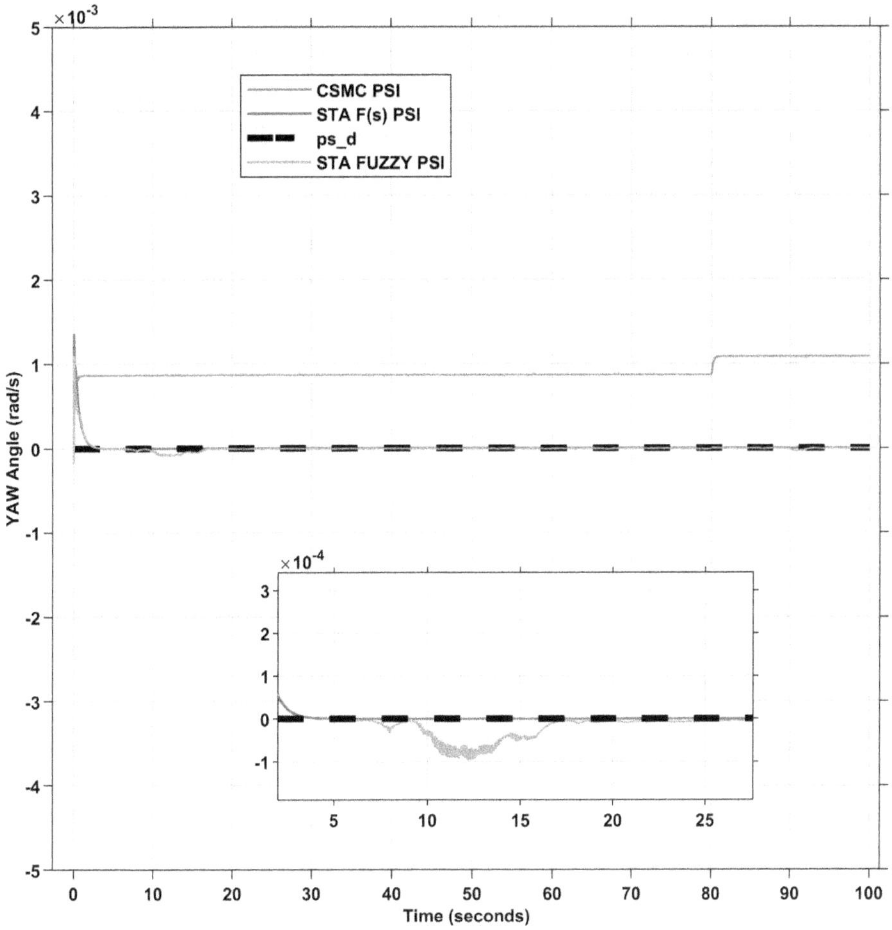

6. CONCLUSION

This work designs a non-linear control law, SMC-ASTA and illustrates its performance for the altitude and attitude control of a quadrotor. To evaluate the proposed control law a comprehensive comparative analysis is conducted with the conventional SMC under the presence of external disturbances, actuator saturation and varying model parameters (e.g., payload variation). The results clearly demonstrate that the proposed SMC-ASTA law outperforms the conventional SMC in terms of trajectory tracking, percentage overshoot, response speed, and control effort. From the performance response it is evident that the steady state error is absent. Quantitative performance comparison using performance indices further validate the superiority of the proposed SMC-ASTA control law. The findings of exhaustive simulation studies consistently support the potential of the proposed SMC-ASTA approach in practical applications.

REFERENCES

Anuar, H. A. B., Plestan, F., Chriette, A., & Kermorgant, O. (2022). Super-Twisting Sliding Mode Control with Adaptive Gain of Quadrotor with Rigid Manipulator. *2022 16th International Workshop on Variable Structure Systems (VSS)*, 53–58. 10.1109/VSS57184.2022.9902104

Chen, Q., Tao, M., He, X., & Tao, L. (2021). Fuzzy Adaptive Nonsingular Fixed-Time Attitude Tracking Control of Quadrotor UAVs. *IEEE Transactions on Aerospace and Electronic Systems*, *57*(5), 2864–2877. doi:10.1109/TAES.2021.3067610

Chu, H., Jing, Q., Chang, Z., Shao, Y., Zhang, X., & Mukherjee, M. (2020). Quadrotor Attitude Control via Feedforward All-Coefficient Adaptive Theory. *IEEE Access : Practical Innovations, Open Solutions*, *8*, 116441–116453. doi:10.1109/ACCESS.2020.3002962

El-Sousy, F. F. M., Alattas, K. A., Mofid, O., Mobayen, S., & Fekih, A. (2022). Robust Adaptive Super-Twisting Sliding Mode Stability Control of Underactuated Rotational Inverted Pendulum With Experimental Validation. *IEEE Access : Practical Innovations, Open Solutions*, *10*, 100857–100866. doi:10.1109/ACCESS.2022.3208412

El-Sousy, F. F. M., Amin, M. M., & Mohammed, O. A. (2022). Robust Adaptive Neural Network Tracking Control With Optimized Super-Twisting Sliding-Mode Technique for Induction Motor Drive System. *IEEE Transactions on Industry Applications*, *58*(3), 4134–4157. doi:10.1109/TIA.2022.3160136

Ghadiri, H., Emami, M., & Khodadadi, H. (2021). Adaptive super-twisting non-singular terminal sliding mode control for tracking of quadrotor with bounded disturbances. *Aerospace Science and Technology*, *112*, 106616. doi:10.1016/j.ast.2021.106616

Huang, T., Li, B., Shah, A., Qin, N., & Huang, D. (2019). Fuzzy Sliding Mode Control for a Quadrotor UAV. *2019 IEEE 8th Data Driven Control and Learning Systems Conference (DDCLS)*, 672–677. 10.1109/DDCLS.2019.8908982

Jasim, W., & Gu, D. (2014). H∞ for quadrotor attitude stabilization. *2014 UKACC International Conference on Control (CONTROL)*, 19–24. 10.1109/CONTROL.2014.6915109

Jing, Q., Chang, Z., Chu, H., Shao, Y., & Zhang, X. (2019). Quadrotor Attitude Control Based on Fuzzy Sliding Mode Control Theory. *2019 Chinese Control Conference (CCC)*, 8360–8364. 10.23919/ChiCC.2019.8865754

Li, Z., Zhai, J., & Karimi, H. R. (2021). Adaptive finite-time super-twisting sliding mode control for robotic manipulators with control backlash. *International Journal of Robust and Nonlinear Control*, *31*(17), 8537–8550. doi:10.1002/rnc.5744

Lian, S., Meng, W., Lin, Z., Shao, K., Zheng, J., Li, H., & Lu, R. (2022). Adaptive Attitude Control of a Quadrotor Using Fast Nonsingular Terminal Sliding Mode. *IEEE Transactions on Industrial Electronics*, *69*(2), 1597–1607. doi:10.1109/TIE.2021.3057015

Modi, A., Joshi, N., & Mehta, A. (2023). Robust discrete-time super twisting formation protocol for a 6-DOF Quadcopter swarm. *ISA Transactions*, 177–187. Advance online publication. doi:10.1016/j.isatra.2023.09.029 PMID:37798204

Nakamura, M., Takaya, K., Ohta, H., Shibayama, K., & Kroumov, V. (2019). Quadrotor Modeling and Simulation for Industrial Application. *2019 23rd International Conference on System Theory, Control and Computing (ICSTCC)*, 37–42. 10.1109/ICSTCC.2019.8885708

Obias, K. C. U., Say, M. F. Q., Fernandez, E. A. V., Chua, A. Y., & Sybingco, E. (2019). A Study of the Interaction of Proportional-Integral- Derivative (PID) Control in a Quadcopter Unmanned Aerial Vehicle (UAV) Using Design of Experiment. *2019 IEEE 11th International Conference on Humanoid, Nanotechnology, Information Technology, Communication and Control, Environment, and Management (HNICEM)*, 1–4. 10.1109/HNICEM48295.2019.9072806

Oliva-Palomo, F., Munoz-Vazquez, A. J., Sanchez-Orta, A., Parra-Vega, V., Izaguirre-Espinosa, C., & Castillo, P. (2019). A Fractional Nonlinear PI-Structure Control for Robust Attitude Tracking of Quadrotors. *IEEE Transactions on Aerospace and Electronic Systems*, *55*(6), 2911–2920. doi:10.1109/TAES.2019.2893817

Seeber, R., Horn, M., & Fridman, L. (2018). A Novel Method to Estimate the Reaching Time of the Super-Twisting Algorithm. *IEEE Transactions on Automatic Control*, *63*(12), 4301–4308. doi:10.1109/TAC.2018.2812789

Shao, X., Sun, G., Yao, W., Liu, J., & Wu, L. (2022). Adaptive Sliding Mode Control for Quadrotor UAVs With Input Saturation. *IEEE/ASME Transactions on Mechatronics*, *27*(3), 1498–1509. doi:10.1109/TMECH.2021.3094575

Shehzad, M. F., Bilal, A., & Ahmad, H. (2019). Position & Attitude Control of an Aerial Robot (Quadrotor) With Intelligent PID and State feedback LQR Controller: A Comparative Approach. *2019 16th International Bhurban Conference on Applied Sciences and Technology (IBCAST)*, 340–346. 10.1109/IBCAST.2019.8667170

Tripathi, V. K., Kamath, A. K., Behera, L., Verma, N. K., & Nahavandi, S. (2022). An Adaptive Fast Terminal Sliding-Mode Controller With Power Rate Proportional Reaching Law for Quadrotor Position and Altitude Tracking. *IEEE Transactions on Systems, Man, and Cybernetics. Systems*, *52*(6), 3612–3625. doi:10.1109/TSMC.2021.3072099

Ullah, N., Mehmood, Y., Aslam, J., Ali, A., & Iqbal, J. (2021). UAVs-UGV Leader Follower Formation Using Adaptive Non-Singular Terminal Super Twisting Sliding Mode Control. *IEEE Access : Practical Innovations, Open Solutions*, *9*, 74385–74405. doi:10.1109/ACCESS.2021.3081483

Xu, B., & Lu, X. (2020). An Online Adaptive Control Strategy for Trajectory Tracking of Quadrotors Based on Fuzzy Approximation and Robust Sliding Mode Algorithm. *IEEE Access : Practical Innovations, Open Solutions*, *8*, 215327–215342. doi:10.1109/ACCESS.2020.3039546

Yogi, S. C., Tripathi, V. K., & Behera, L. (2021). Adaptive Integral Sliding Mode Control Using Fully Connected Recurrent Neural Network for Position and Attitude Control of Quadrotor. *IEEE Transactions on Neural Networks and Learning Systems*, *32*(12), 5595–5609. doi:10.1109/TNNLS.2021.3071020 PMID:33881998

Zhai, J., & Li, Z. (2022). Fast-Exponential Sliding Mode Control of Robotic Manipulator With Super-Twisting Method. *IEEE Transactions on Circuits and Wystems. II, Express Briefs*, *69*(2), 489–493. doi:10.1109/TCSII.2021.3081147

APPENDIX

To facilitate a better understanding of the paper's content, this section highlights several key derivations that have been employed throughout the study.

$$r_1 = \frac{I_y - I_z}{I_x} \text{ (A.I)}$$

$$r_2 = \frac{Jr}{I_x} \text{ (A.II)}$$

$$r_3 = \frac{I_z - I_x}{I_y} \text{ (A.III)}$$

$$r_4 = \frac{Jr}{I_y} \text{ (A.IV)}$$

$$r_5 = \frac{I_x - I_y}{I_z} \text{ (A.V)}$$

$$q_1 = \frac{l}{I_x} \text{ (A.VI)}$$

$$q_2 = \frac{l}{I_y} \text{ (A.VII)}$$

$$q_3 = \frac{l}{I_z} \text{ (A.VIII)}$$

Chapter 5
Enhancing Aerospace Engineering Through Synergies of AI and Blockchain Technologies

P. Chitra
GITAM University, Bangalore, India

A. Saleem Raja
University of Technology and Applied Sciences, Shinas, Oman

ABSTRACT

The aerospace industry stands at the forefront of technological advancements, seeking innovative solutions to enhance safety, efficiency, and sustainability. This chapter explores the synergies between artificial intelligence (AI) and blockchain technologies, presenting optimization techniques that promise transformative impacts on aerospace engineering practices. The integration of AI algorithms in aerospace operations spans autonomous systems, predictive maintenance, design optimization, and crew support. Autonomous navigation and control benefit from AI's adaptive decision-making, while predictive maintenance algorithms revolutionize aircraft component upkeep, reducing downtime and operational costs. Furthermore, AI-driven design optimization refines aerodynamics and fuel efficiency, ushering in a new era of aircraft performance. This chapter highlights real-world examples that underscore the importance of data integrity in aerospace, showcasing instances where accurate information is critical for mission success.

DOI: 10.4018/979-8-3693-1491-3.ch005

1. INTRODUCTION

The aerospace industry, characterized by its pursuit of cutting-edge technologies, continuous innovation, and stringent safety standards, is undergoing a transformative phase driven by the convergence of artificial intelligence (AI) and blockchain technologies. This introduction sets the stage for an exploration of how the integration of AI and blockchain optimization techniques is reshaping the landscape of aerospace engineering, introducing novel approaches to enhance operational efficiency, data security, and overall performance (Abdulrahman et al., 2023; Ahmad et al., 2021; Walthall, 2020).

The advent of AI has ushered in a new era of intelligent automation and decision-making capabilities in aerospace operations. From autonomous systems and predictive maintenance to design optimization and crew support, AI is at the forefront of revolutionizing how aircraft are designed, operated, and maintained. Concurrently, blockchain technology has emerged as a secure and transparent means of managing data, particularly in complex supply chain ecosystems. The immutable nature of blockchain ensures data integrity, while smart contracts automate and streamline various aspects of aerospace transactions.

This chapter delves into the symbiotic relationship between AI and blockchain, examining their collaborative potential in optimizing aerospace engineering practices. Real-world examples underscore the critical importance of data integrity in aerospace, showcasing instances where accurate information is paramount for mission success and safety.

Figure 1. Aviation in artificial intelligence

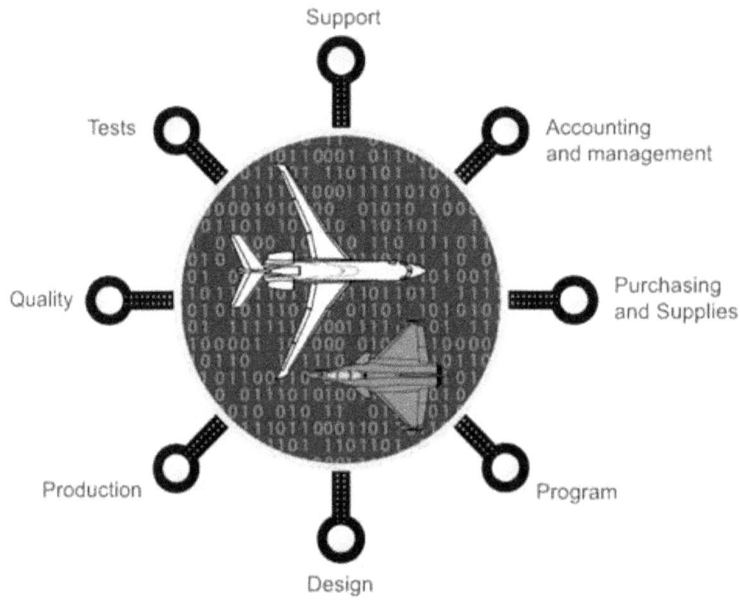

As we explore the application of AI algorithms in autonomous navigation, predictive maintenance, and design optimization, the narrative will unveil how these advancements contribute to safer, more efficient, and environmentally conscious aerospace systems. Simultaneously, the integration of blockchain technology promises to fortify data security, introducing decentralized access controls, encrypted transactions, and transparent supply chain traceability.

Looking ahead, the chapter anticipates emerging trends in AI, such as advanced autonomous systems and explainable AI, and their potential impact on aerospace engineering. The discussion also encompasses the role of blockchain in ensuring data transparency and security in the face of evolving challenges, including counterfeit parts and fraudulent transactions.

In essence, this chapter aims to provide a comprehensive understanding of how the collaborative utilization of AI and blockchain optimization techniques is reshaping the aerospace industry. From securing supply chains to advancing autonomous flight systems, the intersection of these technologies holds immense promise for the future of aerospace engineering, fostering a safer, more efficient, and technologically advanced airspace (Li et al., 2020; Shmelova et al., 2020; Singh et al., 2022; Zavareh et al., 2021).

2. OVERVIEW ON AEROSPACE ENGINEERING USING AI AND BLOCKCHAIN TECHNOLOGY

The integration of artificial intelligence (AI) and blockchain optimization techniques is reshaping the landscape of aerospace engineering, ushering in a new era of innovation, efficiency, and data security. This overview provides a high-level perspective on the collaborative impact of AI and blockchain technologies in the aerospace industry.

2.1. AI Revolutionizing Aerospace Operations

Autonomous Systems: AI algorithms are driving advancements in autonomous systems, enhancing the navigation, control, and decision-making capabilities of aerospace vehicles. From unmanned aerial vehicles (UAVs) to autonomous spacecraft, AI is at the forefront of transforming aerospace operations (Kumar et al., 2022; Nguyen et al., 2021; Sánchez Pérez et al., 2020).

Predictive Maintenance: AI is optimizing aircraft maintenance through predictive algorithms that analyze sensor data to forecast potential component failures. This approach minimizes downtime, extends the lifespan of critical components, and leads to cost savings for aerospace operators.

Figure 2. Artificial intelligence in aviation market (2021-2030)

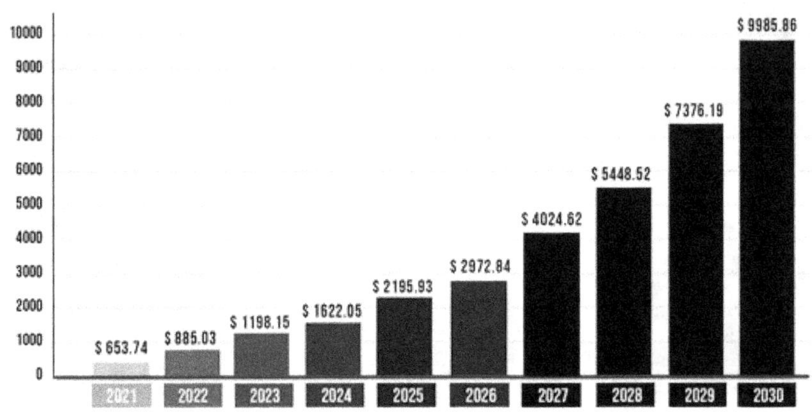

Design Optimization: Generative design algorithms powered by AI are revolutionizing aircraft design. By optimizing aerodynamics, fuel efficiency, and structural integrity, AI-driven design processes contribute to the development of high-performance and environmentally sustainable aircraft.

Crew Support: AI applications support flight crews by providing real-time insights, automating routine tasks, and enhancing overall safety. Virtual co-pilots and decision support systems leverage AI to improve decision-making in various flight scenarios.

2.2. Blockchain Securing Aerospace Data

Supply Chain Traceability: Blockchain ensures end-to-end traceability in the aerospace supply chain, mitigating risks associated with counterfeit parts and enhancing transparency. Each transaction is securely recorded on an immutable ledger, providing a tamper-proof record of the supply chain journey.

Smart Contracts in Aerospace Transactions: Smart contracts on the blockchain automate and secure various aerospace transactions, including leasing, financing, and maintenance records. This streamlines processes, reduces paperwork, and enhances transparency in financial transactions related to aerospace assets.

Figure 3. Visualization of aerospace engineering using AI and blockchain technology

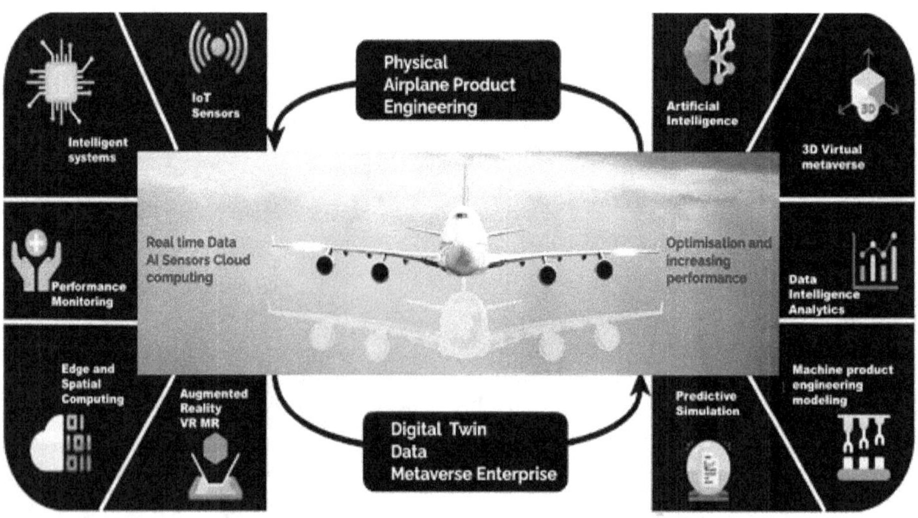

Digital Twins and Blockchain Integration: Blockchain enhances the security and transparency of digital twin data, which represents virtual counterparts of physical aerospace components. The decentralized and immutable nature of blockchain ensures the integrity of digital twin information (Kumar Nimbark, 2023; Manoharan et al., 2022; Oche et al., 2021).

Secure Supply Chain Management: Blockchain technology introduces decentralized access controls, encrypted transactions, and secure digital identities in aerospace supply chain management. This fortifies data security, reducing the risk of unauthorized access or manipulation of sensitive aerospace information.

2.3. Real-World Examples and Importance of Data Integrity

Examining real-world instances, such as the Mars Climate Orbiter failure and Boeing 737 MAX incidents, underscores the critical importance of data integrity in aerospace. Accurate and reliable information, from units of measurement to sensor data, is imperative for the safety and success of aerospace missions.

2.4. Emerging Trends and Future Developments

Anticipating future trends, the overview explores emerging AI trends, including advanced autonomous systems and explainable AI. Additionally, it discusses the integration of quantum computing with AI and blockchain, opening new possibilities for solving complex aerospace engineering challenges.

In summary, the collaborative integration of AI and blockchain optimization techniques is transforming aerospace engineering across various domains. From enhancing operational efficiency and safety to securing and transparently managing data, this overview sets the stage for a detailed exploration of the dynamic and evolving intersection of AI and blockchain in the aerospace industry.

3. CHALLENGES IN AEROSPACE ENGINEERING USING AI TECHNIQUES

Rapid Technological Changes:

The aerospace industry faces challenges in keeping pace with rapidly evolving technologies, including advancements in materials, propulsion systems, and avionics.

Cost Management:

Developing and manufacturing aircraft involves substantial costs. Aerospace companies continually grapple with the need to balance innovation with cost efficiency.

Regulatory Compliance:

Stringent safety and regulatory standards require meticulous compliance, adding complexity and time to the development and certification processes.

Environmental Impact:

Increasing environmental concerns necessitate a shift towards more sustainable practices. Reducing carbon emissions and noise pollution are critical challenges for the industry.

Global Supply Chain Disruptions:

The aerospace industry is highly dependent on a global supply chain. Disruptions, such as geopolitical events or pandemics, can lead to delays and increased costs.

Cybersecurity Threats:

As aircraft become more connected, the industry faces growing cybersecurity threats. Ensuring the security of avionics systems and data is a paramount concern.

4. OPPORTUNITIES IN AEROSPACE ENGINEERING USING AI

a) Advancements in Materials and Manufacturing:

Opportunities arise from the development of lightweight materials, additive manufacturing, and innovative production processes, leading to more fuel-efficient and cost-effective aircraft.

b) Urban Air Mobility (UAM):

The emergence of UAM presents new opportunities for short-distance, on-demand air transportation, fostering innovation in electric propulsion and autonomous systems.

c) Data-Driven Decision Making:

Big data analytics and artificial intelligence enable more informed decision-making, from predictive maintenance to optimizing flight routes, improving efficiency and reducing costs.

d) Sustainable Aviation:

The industry can capitalize on the demand for sustainable aviation solutions, including electric and hybrid-electric propulsion, biofuels, and improved aerodynamics.

e) Space Exploration and Tourism:

Growing interest in space exploration and tourism opens up opportunities for aerospace companies to contribute to the development of space vehicles, satellite technologies, and space tourism initiatives.

f) Collaboration and Partnerships:

Collaboration between aerospace companies, governments, and startups can lead to innovative solutions. Partnerships in research and development can help address challenges more effectively.

g) Military and Defense Contracts:

Defense contracts provide a stable source of revenue for many aerospace companies. Continued innovation in military aviation technology presents lucrative opportunities.

h) Digital Transformation:

The adoption of digital technologies, such as blockchain and the Internet of Things (IoT), offers opportunities for streamlining operations, enhancing supply chain transparency, and improving overall efficiency.

Figure 4. Advantages of artificial intelligence in aerospace engineering

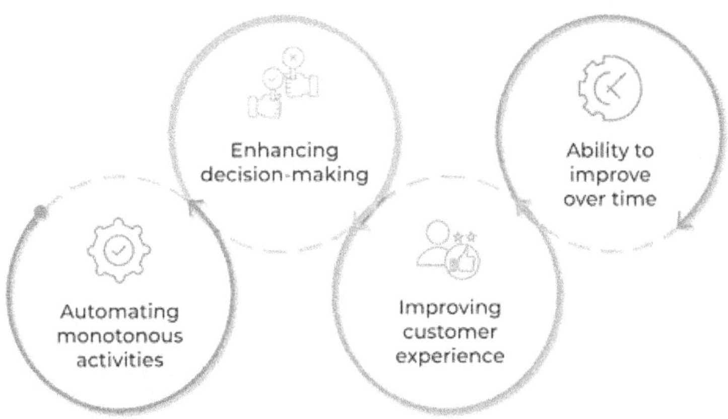

Balancing these challenges and opportunities requires a strategic approach, ongoing innovation, and collaboration across the aerospace ecosystem. The industry's ability to navigate these complexities will shape its future success and contribution to global transportation and technological advancement (Efthymiou et al., 2022; Gupta et al., 2021; Lee et al., 2020).

Purpose of the chapter: Exploring how the integration of AI and blockchain can optimize various facets of aerospace engineering.

5. AI APPLICATIONS IN AEROSPACE ENGINEERING

AI applications in aerospace engineering are diverse, ranging from autonomous systems and predictive maintenance to design optimization and crew support. Here are some key AI applications in the aerospace industry:

a) Autonomous Systems:

Autonomous Navigation: AI algorithms enable autonomous navigation for aircraft and drones, including take-off, landing, and route optimization.

Collision Avoidance: Machine learning algorithms help aircraft identify and avoid obstacles in real-time, enhancing safety during flight.

b) Predictive Maintenance:

AI is used to predict when components of an aircraft are likely to fail based on data from sensors and historical performance metrics. This enables proactive maintenance, reducing downtime and increasing overall efficiency.

c) Design Optimization:

Machine learning algorithms analyze vast datasets to optimize aircraft design. This includes improving aerodynamics, reducing fuel consumption, and enhancing overall performance.

d) Crew Support:

AI systems support flight crews through various applications, including decision support systems and natural language interfaces. Virtual co-pilots powered by AI can assist in decision-making during flight.

e) Supply Chain Optimization:

AI algorithms are employed to optimize the aerospace supply chain. This includes predicting demand, managing inventory, and streamlining logistics for improved efficiency and cost reduction.

f) Flight Route Optimization:

AI is used to analyze real-time weather data, air traffic conditions, and other variables to optimize flight routes. This improves fuel efficiency and reduces flight time.

g) Air Traffic Management:

AI contributes to air traffic management by enhancing coordination and communication between aircraft. This includes automated systems for managing air traffic flow and reducing congestion.

h) Fault Detection and Diagnostics:

Machine learning algorithms analyze sensor data to detect faults and anomalies in aircraft systems. This helps in identifying potential issues early on, preventing failures during flight.

i) Natural Language Processing (NLP):

NLP is applied to improve communication between ground control, pilots, and AI systems. This facilitates more efficient and natural interactions, particularly in complex and dynamic situations.

j) Simulations and Training:

AI is used to create realistic simulations for pilot training. These simulations help pilots experience various scenarios, enhancing their skills and decision-making abilities.

Image and Video Analysis:

AI-based image and video analysis is employed for tasks such as object recognition, satellite imagery interpretation, and surveillance in aerospace applications.

k) Cybersecurity:

AI is utilized to enhance cybersecurity in aerospace systems. It helps detect and prevent cyber threats, ensuring the integrity and security of critical avionics systems.

These applications showcase the versatility of AI in addressing challenges and improving various aspects of aerospace engineering, ultimately contributing to safer, more efficient, and technologically advanced aviation.

6. AUTONOMOUS SYSTEMS: OVERVIEW OF AI ALGORITHMS FOR AUTONOMOUS NAVIGATION AND CONTROL

AI algorithms for autonomous navigation and control play a pivotal role in enabling vehicles, drones, and robots to operate independently without human intervention. These algorithms leverage various techniques from artificial intelligence, machine learning, and robotics to make decisions based on sensor inputs and environmental data (Yadav et al., 2022). Here's an overview of key AI algorithms used for autonomous navigation and control:

a) Path Planning Algorithms:

A Algorithm: Widely used for finding the shortest path between points in a graph. It is suitable for navigating environments with known structures.

D Algorithm: Dynamically adjusts paths in real-time, making it suitable for environments where the layout may change.

Rapidly Exploring Random Trees (RRT):

Builds a tree structure to explore the search space efficiently, often used in dynamic and complex environments.

b) Localization Algorithms:

Simultaneous Localization and Mapping (SLAM):

Enables a system to create a map of its environment while simultaneously determining its location within that environment.

Extended Kalman Filter (EKF):

Fusion of sensor data with predictions to estimate the vehicle's position, especially useful in environments with noise.

c) Control Algorithms:

Proportional-Integral-Derivative (PID) Controller:

A classic control algorithm that adjusts the system's output based on the error, integral, and derivative terms. It is widely used in controlling the stability of autonomous vehicles.

Model Predictive Control (MPC):

Optimizes control inputs over a prediction horizon, considering the system dynamics and constraints. It is effective for systems with complex dynamics.

d) Obstacle Avoidance Algorithms:

Potential Fields:

Creates a virtual force field around obstacles, and the vehicle navigates by moving along the steepest gradient toward the goal while avoiding obstacles.

Reactive Algorithms:

Make decisions based on immediate sensor inputs, reacting quickly to obstacles or changes in the environment.

e) Machine Learning Algorithms:

Reinforcement Learning (RL):

Trains agents to make sequential decisions by learning from trial and error. RL is used for decision-making in dynamic and unpredictable environments.

Deep Neural Networks (DNN):

Utilized for perception tasks such as image recognition and object detection, providing crucial inputs for navigation decisions.

Convolutional Neural Networks (CNN):

Particularly useful for processing visual data, such as camera inputs, for object recognition and scene understanding.

f) Sensor Fusion:

Kalman Filters and Bayesian Filters:

Combine information from multiple sensors, such as GPS, IMU, lidar, and radar, to improve accuracy in estimating the vehicle's state.

g) Behavioral Planning:

Finite State Machines (FSM): Define different states and transitions between states to represent the vehicle's behavior in different scenarios.

Decision Trees: Make decisions based on a series of logical conditions, determining the appropriate action in response to specific situations.

h) Natural Language Processing (NLP):

Human-Machine Interface (HMI): Allows for communication between autonomous systems and human operators through natural language interfaces.

These algorithms are often integrated into a comprehensive autonomous system that combines perception, decision-making, and control to navigate and operate in complex and dynamic environments. The choice of algorithms depends on the specific requirements and challenges of the autonomous navigation task at hand (Chang et al., 2019; Khunaifi et al., 2023; Raoofi & Yasar, 2023).

7. CASE STUDIES ON AUTONOMOUS TAKE-OFF, LANDING, AND ROUTE OPTIMIZATION

While specific case studies may be limited due to the proprietary nature of some developments in autonomous aerospace technology, there are notable examples and advancements in autonomous take-off, landing, and route optimization (Raj et al., 2022). Here are a few instances:

a) Autonomous Take-Off and Landing:

Boeing's Autonomous Passenger Air Vehicle (PAV):

Overview: Boeing NeXt's Passenger Air Vehicle (PAV) project aimed to develop an autonomous electric vertical takeoff and landing (eVTOL) aircraft for urban air mobility.

Case Study:

Boeing's PAV successfully demonstrated autonomous take-off, flight, and landing capabilities. The aircraft utilized a range of sensors, including lidar and cameras, along with sophisticated AI algorithms to navigate and make real-time decisions during the flight.

b) Autonomous Route Optimization:

Airbus Skyways Project:

Overview: The Airbus Skyways project focused on developing autonomous parcel delivery drones for urban environments. It aimed to optimize routes for efficient and safe deliveries.

Case Study:

In a pilot project conducted in Singapore, Airbus tested autonomous drones that could navigate through urban landscapes to deliver packages. The system used

AI algorithms to optimize routes based on real-time data, avoiding obstacles and ensuring timely deliveries.

c) Autonomous Take-Off, Landing, and Route Optimization:

NASA's Unmanned Aircraft Systems Traffic Management (UTM):

Overview: NASA's UTM project aimed to develop a system for managing drone traffic in low-altitude airspace. It involves autonomous capabilities for take-off, landing, and route optimization.

Case Study:

NASA has conducted various tests and demonstrations of UTM technologies. The system integrates AI algorithms to optimize drone routes, ensuring safe separation between vehicles and avoiding restricted areas. This project showcases a holistic approach to autonomous operations in airspace.

d) Autonomous Landing on an Aircraft Carrier:

X-47B Unmanned Combat Air System (UCAS):

Overview: The X-47B UCAS, developed by Northrop Grumman for the U.S. Navy, is an autonomous drone designed for carrier-based operations.

Case Study: The X-47B demonstrated the ability to autonomously take off, navigate, and land on an aircraft carrier. This marked a significant advancement in autonomous aviation technology, showcasing the feasibility of unmanned systems for complex and dynamic operations.

e) Autonomous Route Optimization for Commercial Flights:

EasyJet and Wright Electric:

Overview: EasyJet collaborated with Wright Electric to explore the development of electric aircraft for short-haul flights, incorporating autonomous features for route optimization.

Case Study: While in the early stages, this collaboration aimed to develop electric aircraft that could autonomously optimize flight routes, reduce fuel consumption, and enhance overall efficiency for short-haul commercial flights. The project reflects a forward-looking approach to integrating autonomous technologies into mainstream aviation.

These case studies highlight the progress and ongoing efforts in the field of autonomous aerospace operations, showcasing successful implementations of autonomous take-off, landing, and route optimization across various applications, from urban air mobility to military and commercial aviation.

8. PREDICTIVE MAINTENANCE

Predictive maintenance is an advanced maintenance strategy that uses data, analytics, and machine learning to predict when equipment or machinery is likely to fail. By monitoring the condition of equipment in real-time, organizations can schedule maintenance activities precisely when they are needed, minimizing downtime and reducing unnecessary maintenance costs. Here's an overview of predictive maintenance:

a) Key Components of Predictive Maintenance:

Sensor Data: Collecting real-time data from sensors attached to equipment or machinery. These sensors monitor various parameters such as temperature, vibration, pressure, and other relevant indicators.

Data Processing and Analysis: Using machine learning algorithms to analyze the collected sensor data. These algorithms identify patterns, anomalies, and trends that can indicate potential issues or failure risks.

b) Predictive Models:

Developing predictive models based on historical data and machine learning algorithms. These models forecast when equipment is likely to fail, providing insights into the remaining useful life of critical components.

Condition Monitoring:

Continuously monitoring the condition of equipment to detect subtle changes that may indicate impending failures. This real-time monitoring allows for proactive decision-making.

Maintenance Triggers:

Establishing predefined thresholds and triggers based on predictive models. When certain conditions or patterns are identified, maintenance alerts are generated, signaling that maintenance is recommended.

c) Integration with Enterprise Systems:

Integrating predictive maintenance systems with other enterprise systems, such as Enterprise Resource Planning (ERP) or Asset Management Systems, to streamline workflows and enhance overall efficiency.

d) Benefits of Predictive Maintenance:

Reduced Downtime: By identifying potential issues before they lead to equipment failure, predictive maintenance helps schedule maintenance activities during planned downtime, minimizing the impact on operations.

Cost Savings: Targeted maintenance activities reduce the need for unnecessary and costly preventive maintenance or emergency repairs. This leads to cost savings in terms of labor, materials, and equipment replacement.

Extended Equipment Life: Predictive maintenance enables organizations to maximize the lifespan of their equipment by addressing issues before they escalate, ultimately improving the reliability and longevity of assets.

Improved Safety: Proactively addressing equipment issues enhances overall safety by reducing the likelihood of unexpected failures that could result in accidents or injuries.

Data-Driven Decision-Making: Leveraging data and analytics allows organizations to make informed decisions regarding maintenance strategies, resource allocation, and equipment upgrades.

Optimized Maintenance Schedules: Predictive maintenance optimizes maintenance schedules based on the actual condition of equipment, preventing over-maintenance or under-maintenance scenarios.

9. REAL-WORLD APPLICATIONS OF AI AND BLOCKCHAIN TECHNOLOGY

Aircraft Maintenance: Airlines use predictive maintenance for aircraft engines, avionics systems, and other critical components to ensure safe and efficient operations.

Manufacturing Industry: Predictive maintenance is widely employed in manufacturing to monitor the health of machinery, reducing unplanned downtime and improving overall production efficiency.

Energy Sector: Power plants utilize predictive maintenance to monitor the condition of turbines, generators, and other critical equipment, ensuring continuous and reliable energy production.

Railway Maintenance: Rail operators apply predictive maintenance to monitor the condition of tracks, locomotives, and rolling stock, optimizing maintenance schedules and minimizing service disruptions.

Predictive maintenance represents a shift from traditional reactive and preventive maintenance approaches, offering organizations a more efficient and cost-effective way to manage their assets and infrastructure.

10. EXPLANATION OF AI-DRIVEN PREDICTIVE MAINTENANCE FOR AIRCRAFT COMPONENTS

AI-driven predictive maintenance for aircraft components involves leveraging artificial intelligence (AI) and machine learning (ML) techniques to analyze data from various sensors and systems on an aircraft. The goal is to predict when specific components are likely to fail or require maintenance before the failure occurs. This proactive approach helps airlines and maintenance teams schedule maintenance activities more efficiently, reducing downtime, minimizing disruptions to flight schedules, and optimizing overall operational costs (Shmelova et al., 2019).

Here's a step-by-step explanation of AI-driven predictive maintenance for aircraft components:

1. Data Collection: Sensor Networks: Aircraft are equipped with a multitude of sensors that monitor various parameters such as engine performance, temperature, pressure, vibration, and more. Flight Data Recorders: Utilize data from flight data recorders, which capture a comprehensive set of parameters during every flight.
2. Data Preprocessing: Clean and preprocess the raw data to remove noise and irrelevant information. Align and synchronize data from various sensors to create a unified dataset.
3. Feature Extraction: Identify relevant features within the dataset that provide valuable information about the health and performance of aircraft components. Features could include temperature trends, vibration patterns, or deviations from normal operating conditions.
4. Historical Data Analysis: Use historical maintenance records and failure data to train machine learning models. The models learn patterns and relationships between features and impending failures during the training phase.
5. Machine Learning Model Training: Select appropriate machine learning algorithms, such as regression models, decision trees, or more advanced techniques like neural networks. Train the models on historical data to predict the remaining useful life of aircraft components.
6. Real-Time Monitoring: Implement the trained models in a real-time monitoring system on the aircraft. Continuously collect and analyze data from sensors during flights.
7. Anomaly Detection: Monitor deviations from the expected behavior using the trained machine learning models. Identify anomalies or patterns indicative of potential issues with specific components.
8. Predictive Alerts: Generate predictive alerts or warnings when the AI system predicts that a particular component is likely to fail within a certain timeframe.

These alerts are sent to maintenance crews, allowing them to plan and schedule maintenance activities proactively.

9. Decision Support System: Provide actionable insights to maintenance teams through a decision support system.

Include recommendations on the type of maintenance needed, whether it's a simple inspection, repair, or replacement of a component.

10. Continuous Learning: Implement a feedback loop for continuous learning. As more data becomes available, update and retrain the machine learning models to improve accuracy and adapt to changing conditions.

11. BENEFITS OF AI-DRIVEN PREDICTIVE MAINTENANCE FOR AIRCRAFT COMPONENTS

Cost Savings: Reduces unnecessary maintenance costs by focusing on components that genuinely need attention.

Downtime Reduction: Minimizes unplanned downtime by addressing potential issues before they lead to component failures.

Optimized Maintenance Scheduling: Allows for more efficient planning of maintenance activities, optimizing resources and minimizing disruptions to flight schedules.

Improved Safety: Enhances overall safety by identifying and addressing potential issues before they pose a risk to flight operations.

Extended Component Life: Maximizes the lifespan of aircraft components by addressing issues proactively, reducing the need for premature replacements.

AI-driven predictive maintenance is a critical component of modern aviation maintenance strategies, providing a data-driven and proactive approach to ensuring the safety, reliability, and efficiency of aircraft operations.

12. REAL-WORLD EXAMPLES SHOWCASING THE IMPACT ON DOWNTIME AND EFFICIENCY

Several real-world examples demonstrate the significant impact of AI-driven predictive maintenance on reducing downtime and improving operational efficiency in the aerospace industry. These examples showcase how airlines and maintenance organizations have leveraged predictive maintenance to optimize their operations:

1. Rolls-Royce's TotalCare Program:

Industry: Aviation (Engines)

Case Study: Rolls-Royce, a major aircraft engine manufacturer, implemented the TotalCare program. This program utilizes data analytics and predictive maintenance to monitor the health of engines in real-time.

By analyzing sensor data and performance metrics, Rolls-Royce can predict potential issues before they lead to engine failures. This proactive approach reduces unplanned maintenance events, minimizing downtime for aircraft operators.

The TotalCare program has been successful in optimizing engine maintenance, improving efficiency, and providing cost predictability for airlines.

2. Lufthansa Technik's AVIATAR Platform:

Industry: Aviation (MRO Services)

Case Study: Lufthansa Technik developed the AVIATAR platform, which integrates various digital services, including predictive maintenance capabilities.

AVIATAR uses AI algorithms to analyze aircraft health data, predict component failures, and recommend maintenance actions. This allows airlines to schedule maintenance during planned downtimes, reducing operational disruptions.

The platform has demonstrated success in improving aircraft availability, reducing maintenance costs, and enhancing overall operational efficiency for its users.

3. Delta TechOps Predictive Maintenance Program:

Industry: Aviation (Airline Maintenance)

Case Study: Delta Air Lines implemented a predictive maintenance program in collaboration with its maintenance division, TechOps.

By leveraging data analytics and machine learning, Delta TechOps predicts component failures on aircraft, allowing for proactive maintenance actions. This has led to a reduction in unscheduled maintenance events and increased aircraft availability.

The program has contributed to cost savings and improved on-time performance for Delta Air Lines.

4. Airbus's Skywise Predictive Maintenance Platform:

Industry: Aviation (Aircraft Manufacturing)

Case Study: Airbus developed the Skywise platform, which includes predictive maintenance capabilities for airlines and operators.

Skywise aggregates and analyzes aircraft operational data to predict potential issues and recommend maintenance actions. This proactive approach enables airlines to avoid unexpected disruptions and optimize maintenance schedules.

Airlines using Skywise have reported improvements in fleet reliability, reduced maintenance costs, and enhanced operational efficiency.

5. Boeing's Maintenance Performance Toolbox:

Industry: Aviation (Aircraft Manufacturing)

Case Study: Boeing offers the Maintenance Performance Toolbox, a suite of digital tools that includes predictive maintenance solutions.

By utilizing machine learning algorithms, Boeing's toolbox predicts potential failures in aircraft systems and components. This enables airlines to plan maintenance activities in advance, minimizing downtime and improving fleet reliability. The toolbox has contributed to increased operational efficiency and cost savings for airlines using Boeing aircraft.

These real-world examples highlight the tangible benefits of AI-driven predictive maintenance in the aerospace industry, including reduced downtime, improved operational efficiency, and cost savings for both aircraft manufacturers and operators. As the industry continues to embrace digital transformation, predictive maintenance is becoming an integral part of modern aviation maintenance strategies.

13. CONCLUSION

The integration of artificial intelligence (AI) and blockchain optimization techniques marks a significant paradigm shift in aerospace engineering, bringing forth a convergence of innovation, efficiency, and enhanced data security. In conclusion, the collaborative efforts between AI and blockchain technologies represent a pivotal moment in the evolution of aerospace engineering. The industry is navigating a path towards increased efficiency, heightened data security, and a future where intelligent systems work in tandem with secure and transparent data management. As the journey unfolds, the aerospace sector stands poised for continued advancements, leveraging the transformative potential of AI and blockchain to reach new heights of safety, efficiency, and technological excellence.

REFERENCES

Abdulrahman, Y., Arnautović, E., Parezanović, V., & Svetinovic, D. (2023). AI and Blockchain Synergy in Aerospace Engineering: An Impact Survey on Operational Efficiency and Technological Challenges. *IEEE Access : Practical Innovations, Open Solutions*, *11*, 87790–87804. doi:10.1109/ACCESS.2023.3305325

Ahmad, R. W., Hasan, H., Yaqoob, I., Salah, K., Jayaraman, R., & Omar, M. (2021). Blockchain for aerospace and defense: Opportunities and open research challenges. *Computers & Industrial Engineering*, *151*, 106982. doi:10.1016/j.cie.2020.106982

Chang, S., Wang, Z., Wang, Y., Tang, J., & Jiang, X. (2019, August). Enabling technologies and platforms to aid digitalization of commercial aviation support, maintenance and health management. In *2019 International Conference on Quality, Reliability, Risk, Maintenance, and Safety Engineering (QR2MSE)* (pp. 926-932). IEEE. 10.1109/QR2MSE46217.2019.9021222

Efthymiou, M., McCarthy, K., Markou, C., & O'Connell, J. F. (2022). An exploratory research on blockchain in aviation: The case of maintenance, repair and overhaul (MRO) organizations. *Sustainability (Basel)*, *14*(5), 2643. doi:10.3390/su14052643

Gupta, R., Kumari, A., & Tanwar, S. (2021). Fusion of blockchain and artificial intelligence for secure drone networking underlying 5G communications. *Transactions on Emerging Telecommunications Technologies*, *32*(1), e4176. doi:10.1002/ett.4176

Khunaifi, A., Wijananto, F., Handoyo, K. M., Juliansyah, M. R., Imdad, M. T., Wibowo, M., ... Arief, M. (2023, February). The Utilization of Blockchain Technology in The Airport Industry: A Review. In *Proceeding of International Conference on Information Science and Technology Innovation (ICoSTEC)* (Vol. 2, No. 1, pp. 145-151). 10.35842/icostec.v2i1.53

Kumar, G. R., Rencher, R., Fabre, C., Budeanu, D., Markou, C., Jones, K., ... Veluri, S. (2022). Best Practices in Establishing Business Case for Implementing Blockchain Solution in Aerospace. *SAE International Journal of Advances and Current Practices in Mobility, 4*, 1104-1114.

Kumar Nimbark, H. (2023). Development of Algorithm for Blockchain and Artificial Intelligence Collaboration to Identify Travel Path of UAVs. *Tuijin Jishu/Journal of Propulsion Technology, 44*(3), 660-666.

Lee, C. K., Huo, Y. Z., Zhang, S. Z., & Ng, K. K. (2020). Design of a smart manufacturing system with the application of multi-access edge computing and blockchain technology. *IEEE Access : Practical Innovations, Open Solutions*, *8*, 28659–28667. doi:10.1109/ACCESS.2020.2972284

Li, W., Su, Z., Li, R., Zhang, K., & Wang, Y. (2020). Blockchain-based data security for artificial intelligence applications in 6G networks. *IEEE Network*, *34*(6), 31–37. doi:10.1109/MNET.021.1900629

Manoharan, H., Teekaraman, Y., Kuppusamy, R., Kaliyan, N., & Thelkar, A. R. (2022). Examining the effect of cyber twin and blockchain technologies for industrial applications using AI. *Mathematical Problems in Engineering*, *2022*, 2022. doi:10.1155/2022/3048038

Nguyen, T. A., Kaliappan, V. K., Jeon, S., Jeon, K. S., Lee, J. W., & Min, D. (2021, November). Blockchain Empowered Federated Learning with Edge Computing for Digital Twin Systems in Urban Air Mobility. In *Asia-Pacific International Symposium on Aerospace Technology* (pp. 935-950). Singapore: Springer Nature Singapore.

Oche, P. A., Ewa, G. A., & Ibekwe, N. (2021). Applications and challenges of artificial intelligence in space missions. *IEEE Access : Practical Innovations, Open Solutions*, 1. doi:10.1109/ACCESS.2021.3132500

Raj, P., Nagarajan, G., & Minu, R. I. (Eds.). (2022). *Applied Edge AI: Concepts, Platforms, and Industry Use Cases*. CRC Press.

Raoofi, T., & Yasar, S. (2023). Analysis of frontier digital technologies in continuing airworthiness management frameworks and applications. *Aircraft Engineering and Aerospace Technology*, *95*(10), 1669–1677. doi:10.1108/AEAT-06-2022-0166

Sánchez Pérez, A. M., Tarifa Fernández, J., & Cruz Rambaud, S. (2020). Assessing blockchain investments through the learning option: An application to the automotive and aerospace industry. *Mathematics*, *8*(12), 2213. doi:10.3390/math8122213

Shmelova, T., Sikirda, Y., & Sterenharz, A. (Eds.). (2019). *Handbook of Research on Artificial intelligence applications in the aviation and aerospace industries*. IGI Global.

Shmelova, T., Sterenharz, A., & Dolgikh, S. (2020). Artificial intelligence in aviation industries: methodologies, education, applications, and opportunities. In *Handbook of research on artificial intelligence applications in the aviation and aerospace industries* (pp. 1–35). IGI Global. doi:10.4018/978-1-7998-1415-3.ch001

Singh, P., Elmi, Z., Lau, Y. Y., Borowska-Stefańska, M., Wiśniewski, S., & Dulebenets, M. A. (2022). Blockchain and AI technology convergence: Applications in transportation systems. *Vehicular Communications*, *38*, 100521. doi:10.1016/j.vehcom.2022.100521

Walthall, R. (2020). *Unsettled Topics Concerning Adopting Blockchain Technology in Aerospace* (No. EPR2020021). SAE Technical Paper.

Yadav, J. K., Verma, D. C., Jangirala, S., Srivastava, S. K., & Aman, M. N. (2022). Blockchain for aviation industry: Applications and used cases. In *ICT Analysis and Applications* (pp. 475–486). Springer Singapore. doi:10.1007/978-981-16-5655-2_46

Zavareh, B., Foroozan, H., Gheisarnejad, M., & Khooban, M. H. (2021). New trends on digital twin-based blockchain technology in zero-emission ship applications. *Naval Engineers Journal*, *133*(3), 115–135.

Chapter 6
Investigation of Aviation Sector in Industry 5.0

G. Prasad
https://orcid.org/0000-0002-5709-9182
Chandigarh University, India

ABSTRACT

Industry 5.0 concepts and technologies are increasingly being adopted in the aviation sector, revolutionizing the way aircraft are designed, manufactured, maintained, and operated. Industry 5.0 leverages digital twin technology, creating virtual replicas of physical aircraft, which can be used for real-time monitoring, testing, and design optimization. Engineers and data scientists work together to create and manage these digital twins. Robotics and automation are used for precision manufacturing and assembly processes. Collaborative robots (cobots) are employed to work alongside human technicians, ensuring safer and more efficient assembly.

INTRODUCTION

Advanced Automation and Robotics, Artificial Intelligence (AI) in Aviation, Data Integration and Connectivity, Sustainability and Green Aviation, Supply Chain Digitization, Augmented Reality (AR) and Virtual Reality (VR) are some potential areas in which the concepts of Industry 5.0 could have an impact on the aviation industry. Other potential areas include Supply Chain Digitization.

Manufacturing of aircraft, aircraft maintenance, and possibly even flight operations might all benefit from additional advances in automation and the utilization of robotics in Industry 5.0. Because of this, there is the potential for higher efficiency, decreased costs, and improved safety. Enhanced uses of artificial intelligence could play a

DOI: 10.4018/979-8-3693-1491-3.ch006

significant part in a variety of facets of the aviation industry, including autonomous flight capabilities, predictive maintenance, route optimization, and even air traffic management. It's possible that Industry 5.0 will place an emphasis on the seamless integration of data across the aviation ecosystem. This might involve improving the connectivity between aircraft systems, ground operations, and other stakeholders, which would result in improved decision-making and increased operational efficiency. The next stage of the aviation industry's industrial evolution may place a greater emphasis on environmentally responsible and sustainable operations. This may involve the development of environmentally friendly technology, such as electric or hybrid propulsion systems, as well as the utilization of resources more efficiently over the whole lifecycle of an aircraft. It's possible that the aircraft supply chain will be fully digital and networked by the time Industry 5.0 arrives. This may result in real-time monitoring of inventory, streamlined logistics, and increased coordination between manufacturers, suppliers, and providers of maintenance services. The combination of augmented reality and virtual reality technology has the potential to improve several facets of the aviation industry, including maintenance, simulation, and inspections.

BACKGROUND OF THE RESEARCH

Aydn & Kahraman, (2021) analysed ion the Aviation 4.0 with theory and applications. Chen et al (2023) developed a concept for human-cyber-physical systems of future wind turbines and energy generation towards Industry 5.0. Elangovan (2021) published a book on Industry 5.0 The future of the industrial economy with the developments for the change in the technology. Ghobakhloo et al (2023) proposed actions and approaches for enabling Industry 5.0-driven sustainable industrial transformation with a strategic roadmap. Jain et al (2021) developed an unmanned aerial vehicle borne secure communication with classification framework for industry 5.0. Javaid & Haleem (2020) conducted a study on critical components of Industry 5.0 towards a successful adoption in the field of manufacturing.

Jafari et al (2022) studied the transformation from Moving Industry 4.0 to Industry 5.0 with respect to the implications for smart logistics. Kavya et al (2022) analysed the influence of Artificial Intelligence, Machine Learning, and Internet of Drones in Medical Applications. Massaro (2021) conducted a research on Advanced Industries: Industry 4.0 to Industry 5.0 Advances. Prasad et al (2023) studied the influence of Machine Learning Technique in Unmanned Aerial Vehicle. Prasad, G. (2023) analysed the Internet of Unmanned Aerial Vehicle (IOU) in Industry 5.0. Sharma & Arya(2022) conducted UAV based long range environment monitoring system with Industry 5.0 perspectives for smart city infrastructure. Thakur & Sehgal (2021) developed an emerging architecture for heterogeneous smart cyber-physical systems for industry 5.0. Zhang et al (2023) studied about new-generation human-centric smart manufacturing in Industry 5.0.

Figure 1. IOT in aviation industry

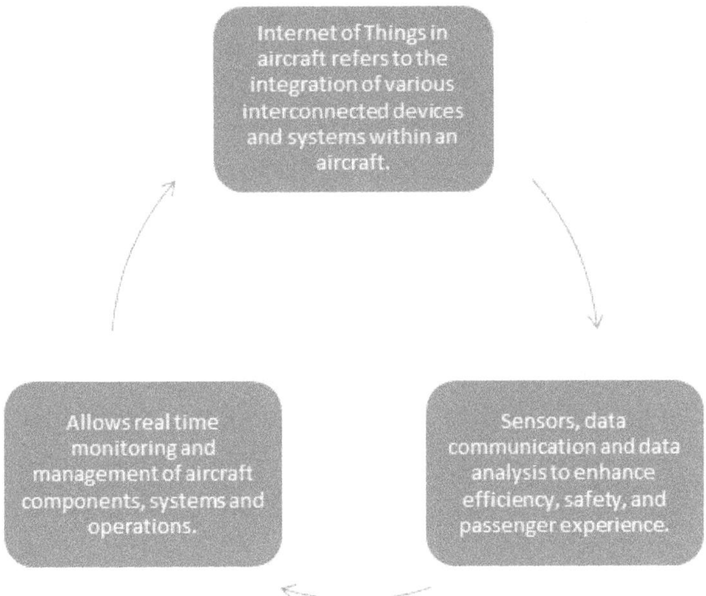

Figure 1 shows the Internet of Things in aircraft refers to the integration of various interconnected devices and systems within an aircraft. Sensors, data communication and data analysis to enhance efficiency, safety, and passenger experience. Allows real time monitoring and management of aircraft components, systems and operations.

INNOVATIVE FORMS OF ROBOTICS AND AUTOMATION

The aircraft industry benefits greatly from the application of cutting-edge automation and robotics technology, which leads to improvements in efficiency, precision, and overall safety. These technologies are having a substantial impact in a number of different fields, some of which are included below. The putting together of airplane parts frequently involves the employment of cutting-edge robotics technology. These robots are able to carry out activities in the manufacturing process like as drilling, riveting, and painting, which results in increased accuracy and productivity. The production of complicated aircraft components is moving more and more toward the usage of additive manufacturing techniques like 3D printing. The process of 3D printing may be automated using robotics, which would improve consistency and reduce the amount of time needed for production. In order to evaluate aircraft surfaces for flaws or damage, modern sensors are installed on drones that are employed for the inspection process. This has the potential to cut down on the amount of time needed for manual inspections while simultaneously increasing safety.

Figure 2. Shows the branches of aviation industry

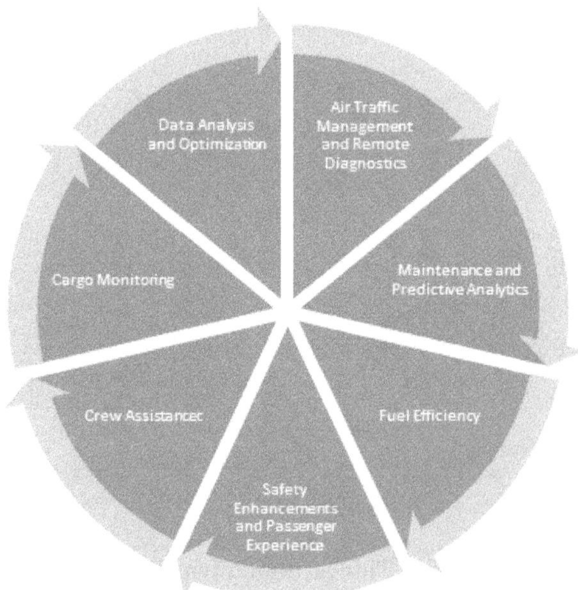

Figure 2 depicts the Air Traffic Management and Remote Diagnostics, Maintenance and Predictive Analytics, Fuel Efficiency, Safety Enhancements and Passenger Experience, Crew Assistance, Cargo Monitoring, Data Analysis and Optimization. Robotics are utilized for a variety of jobs associated with ordinary maintenance, including cleaning, painting, and even more intricate repairs. This helps to reduce the amount of time the aircraft is out of service and ensures that it stays in top shape. UAVs, often known as drones, are utilized widely in the aerospace sector for the purposes of surveillance and monitoring. They are able to do duties such as the inspection of pipelines and infrastructure, as well as inspect broad areas, and deliver data in real time. There are some businesses looking at the possibility of using drones to deliver consumer goods or even components for aerospace vehicles. The use of robotic rovers to investigate the terrain of planets and moons demonstrates the significant contribution made by robotics to the field of space exploration. These rovers are outfitted with a wide variety of tools and devices, allowing scientists to conduct experiments and collect data using the rovers. Automated systems are utilized in the process of deploying satellites as well as in certain instances for the purposes of satellite maintenance and repair missions. The optimization of air routes, the reduction of congestion, and the enhancement of safety can all be achieved through the application of advanced automation in air traffic management systems. This includes the application of artificial intelligence in order to forecast and manage the flow of air traffic. Most contemporary airplanes are fitted with very advanced autopilot systems that enable the pilot to control a wide range of flying characteristics. These systems help to more efficient use of fuel and lower the amount of effort required of the pilot. Robotics and automation play a part in the development of fly-by-wire systems, which allow electronic systems to operate the flight control surfaces of an airplane. These systems were developed in part thanks to robotics and automation. In order to create flight simulators that provide pilots with realistic training environments, robotics is used in the development process. These simulators are able to mimic a wide variety of flight scenarios, including emergencies. The management of supply chains in the aerospace industry makes use of robotics in several applications, such as automated warehouses and logistics systems for effective inventory management and delivery.

THE APPLICATION OF ARTIFICIAL INTELLIGENCE (AI) TO AVIATION

There has been a steady increase in the use of artificial intelligence (AI) across the board in the aviation industry, which has opened up new doors for increased productivity, improved safety, and more informed decision-making. Flight Operations and Planning, Air Traffic Management, Aircraft Systems and Autonomy, Airline Customer Service, Crew Assistance and Training, Security and Threat Detection, Baggage Handling and Logistics are just some of the areas where AI is making a significant influence in the aviation industry.

The most efficient flight paths are determined by AI algorithms after taking into account historical data, current weather conditions, and active air traffic. This could result in savings on fuel, a reduction in pollutants, and improved efficiency in flight planning. By evaluating the data provided by aircraft sensors, artificial intelligence is utilized to forecast when equipment may fail and what maintenance would be required. This makes it easier for airlines to schedule maintenance in advance, which cuts down on unscheduled downtime. AI is used to increase efficiency in the flow of air traffic, control congestion, and make better use of available airspace.

Automated air traffic control systems have the potential to both improve safety and cut down on delays. Collision avoidance systems that are powered by AI analyze the data collected by sensors to make predictions about the likelihood of prospective crashes and either suggest or carry out preventative measures. Artificial intelligence plays a part in the development of improved autopilot systems, which enables more precise control of airplanes. These devices are able to adapt to the constantly shifting environment and improve fuel efficiency. Continued research and development are being done in order to achieve fully or partially self-driving aircraft.

Artificial intelligence is a crucial component that makes it possible for these aircraft to make judgments and navigate without the assistance of humans. Chatbots and virtual assistants powered by AI are increasingly being utilized by airlines to provide customer care. They are able to provide information in real time, aid with bookings, and answer inquiries of a routine nature. Artificial intelligence is used by airlines to evaluate passenger data and preferences, which enables the airlines to provide individualized services and make recommendations. AI technologies provide pilots with information, ideas, and alarms in real time, thereby assisting them in their jobs. These technologies help to improve both the user's awareness of the situation and their ability to make decisions. The incorporation of AI into flight training simulators allows for the creation of more realistic scenarios, better adaptation of training programs to individual performance, and overall improvement in the quality of pilot education. Artificial intelligence is applied to the task of identifying potential security concerns by evaluating the behavior of passengers and crew. This involves keeping an eye out for any strange goings-on and examining the information gathered at the various security checkpoints. Algorithms based on artificial intelligence are able to examine enormous amounts of data in order to forecast prospective security threats or vulnerabilities, which assists authorities in taking preventative actions. The use of AI helps to optimize baggage handling systems, which in turn reduces the likelihood of luggage being lost or delayed. Artificial intelligence is utilized by automated systems to properly track and route baggage. AI is utilized in the aviation industry for the management of its supply chain and logistics, for the purpose of optimizing inventories, and for the purpose of ensuring timely availability of spare parts.

INTEGRATION OF DATA AND NETWORKING CONNECTIVITY

Integration and connection of data are two key factors that provide a significant contribution to improvements in productivity, safety, and decision-making processes within the aviation industry. Data integration and connection have a substantial impact in a number of important areas, including the following: flight operations, air traffic management, passenger services, airport operations, maintenance and logistics, safety and security, regulatory compliance, and training and simulation.

The incorporation of real-time weather information into flight planning systems is made possible by airlines' use of data integration. Because of this, pilots are able to make educated decisions about adjusting their flight paths, reducing their fuel consumption, and avoiding turbulence. Real-time monitoring of the aircraft's health is made possible through the integration of data from onboard sensors. This makes predictive maintenance easier to perform, which in turn decreases downtime and increases overall fleet reliability. Collaborative decision-making is supported by the sharing of data and integration of systems among many stakeholders, such as airports, airlines, and those in charge of air traffic control. This helps to optimize the flow of air traffic, which in turn reduces delays and improves the overall management of airspace. Both pilots and air traffic controllers benefit from an increase in their situational awareness when aircraft equipped with ADS-B transponders transmit their position and other information to other nearby aircraft. while it comes to providing passengers with a smooth experience while purchasing their flights, airlines use integrated booking systems that interact with a variety of platforms, travel agencies, and other third-party services. Connectivity in flight is made possible through the integration of satellite and ground-based communication technologies, which enables passengers to use the internet, make phone calls, and stream entertainment while they are in flight. Integration of data provides assistance for A-CDM programs, which involve the sharing of information across various airport stakeholders for the purpose of improving airport efficiency overall, as well as optimizing ground operations and reducing turnaround times.

The integration of data from different baggage handling systems enables real-time tracking of luggage, which in turn reduces the likelihood of bags being lost or delayed. The integration of data across all stages of the aviation supply chain helps to expedite logistics, enhance inventory management, and guarantee the timely availability of spare parts. Electronic Flight Bags, or EFBs, are integrated systems that give pilots access to real-time data, charts, and manuals. Pilots utilize EFBs. EFBs improve the effectiveness of the cockpit and decrease the dependency on paper documents. Integration of data is essential for security systems that monitor and analyze data gathered from a variety of sources, including as surveillance cameras, access control systems, and passenger screening, in order to guarantee the safety of airports and planes as a whole. Data on threat information is shared between airlines and aviation authorities in order to improve cybersecurity safeguards and reduce the likelihood of potential issues. Integrative systems are used by airlines and aviation organizations to manage and monitor their level of regulatory compliance. This guarantees that safety, security, and operational standards are adhered to at all times. Data integration is applied in training simulators to give realistic situations and simulate diverse settings. This enables pilots and other aviation professionals to improve their skills in a controlled environment while honing their expertise.

ENVIRONMENTAL RESPONSIBILITY AND CLEAN FLYING

As concerns about climate change and the impact humans have on the environment continue to grow, the aviation sector is recognizing the growing importance of sustainability and environmentally conscious activities. Alternative fuels, electric and hybrid propulsion, aircraft design and materials, energy-efficient technologies, carbon offsetting and reduction, operational efficiency, regulatory measures, investment in research and development, public awareness and education are some of the key aspects and initiatives related to sustainability and green aviation.

The production and use of sustainable aviation fuels (SAFs), which are generated from bio-based sources like algae or waste materials, with the goal of lowering aviation's carbon footprint, are now under development and development. Biofuels can be mixed with regular jet fuel, and the resulting mixture can then be used in aircraft engines that are already in use. There is now continuing research and development in the subject of electric propulsion for aircraft of a smaller size. The use of electric propulsion has the potential to drastically cut down on emissions as well as noise pollution. Some businesses are investigating the possibility of developing hybrid-electric propulsion systems, which combine conventional and electric power sources to achieve greater fuel efficiency and lessen their negative impact on the environment. When designing airplanes, the utilization of lightweight and composite materials contributes to total weight reduction, which in turn helps increase fuel efficiency. The reduction of drag and the improvement of fuel efficiency are both contributed to by advancements in aerodynamics. In order to improve the performance of the aircraft, winglets and several other design modifications are installed.

Ongoing research is aimed at designing airplane engines that are both more ecologically friendly and efficient in their use of fuel. Technologies such as open-rotor designs and high-bypass turbofans are examples of technologies that strive to improve the efficiency of propulsion. The activities that take place on the ground are a considerable source of emissions. The utilization of electric taxiing systems, in which aircraft taxi utilizing electric power on the ground, might lessen the requirement for the use of main engines when the aircraft is in taxi mode. Airline companies and aviation groups often take part in carbon offset schemes, in which they invest in initiatives that either cut or capture an equivalent quantity of carbon dioxide to balance off their emissions. Some airlines have made the commitment to operate their businesses in a carbon-neutral manner by making investments in carbon offsetting and putting sustainable policies into place. The application of advanced data analytics and artificial intelligence in flight planning helps to optimize routes, which in turn helps to reduce fuel consumption as well as emissions.

Continuous descent approaches are being used more frequently during landings by aircraft because they reduce the amount of fuel burned and the amount of noise produced in comparison to more traditional step-down approaches. The International Civil Aviation Organization (ICAO), among other international bodies, is responsible for determining the emission limits that apply to airplanes. More stringent regulations encourage businesses to create and implement more environmentally friendly technologies. In order to incentivize reductions in emissions from the aviation sector, some regions have already implemented market-based solutions, such as emissions trading or pricing of carbon, while others are considering doing so. Governments and other stakeholders in the aviation industry make investments in research and development to investigate new technologies and developments that have the potential to make aviation more environmentally friendly. Direct flights, fuel-efficient airlines, and individual carbon footprint reduction are some of the sustainable travel practices that are being promoted through public awareness campaigns that are being run by airlines and industry organizations. These campaigns aim to encourage sustainable travel behaviors such as these.

To tackle the environmental issues that are exacerbated by air travel, the aviation industry will need to take a holistic approach to sustainability. This will require a combination of technology innovation, operational efficiency, regulatory measures, and public engagement. As the aviation sector continues to undergo change, there is likely to be an increased emphasis placed on environmentally friendly aviation, along with continued attempts to create and put into practice environmentally friendly practices and technologies.

THE DIGITIZATION OF THE SUPPLY CHAIN

The term "supply chain digitization" refers to the practice of incorporating digital technologies and data-driven procedures within the aerospace industry in order to improve operational effectiveness, supply chain visibility, and teamwork. The following is a list of important issues and efforts connected to the digitalization of supply chains in the aviation industry:

Digital Twin Technology, Blockchain for Transparency, RFID and IoT Integration, Advanced Analytics and Predictive Maintenance, Cloud-Based Collaboration, Automation and Robotics, Supplier Relationship Management (SRM) Systems, Cybersecurity Measures, Paperless Documentation, 3D Printing/Additive Manufacturing, and Regulatory Compliance Platforms are some of the technologies that are expected to be implemented in the near future.

The technology known as "digital twin" is used to make digital copies of real-world assets, such as airplanes or components. This can assist optimize maintenance, monitor component health, and streamline production operations across the supply chain. The implementation of blockchain technology in the supply chain can lead to the creation of transactions that are both safe and transparent. This helps to verify that the data pertaining to the procurement, production, and distribution of aerospace components are accurate. Devices that use radio-frequency identification (RFID) and the Internet of Things (IoT) are connected into supply chains to provide real-time tracking of inventories, components, and equipment. This not only increases visibility but also decreases the likelihood of making mistakes. Utilizing sophisticated analytics, such as predictive maintenance algorithms, helps anticipate maintenance demands and optimizes the supply chain by assuring timely availability of spare parts. This can be accomplished through the utilization of predictive maintenance algorithms. Analytics tools are used to keep track of how well suppliers are performing, how well logistics are working, and how healthy the supply chain as a whole is.

Cloud-based platforms make it possible to centralize data storage, which in turn enables real-time cooperation and the sharing of information among the various stakeholders in the supply chain. These stakeholders include manufacturers, suppliers, and logistics providers. Robotics and other forms of automation are used in warehouses to perform operations such as picking, packaging, and sorting; this helps to increase both the speed at which orders are fulfilled and their accuracy. Robots are utilized in quality control and inspection activities, which helps to guarantee that aerospace components are up to par with rigorous requirements. The use of SRM systems makes it possible for manufacturers and suppliers to work together. These platforms simplify communication, order processing, and the exchange of data, which in turn helps to develop stronger relationships and more effective business dealings. As the amount of data that is stored digitally continues to grow, there has been a corresponding increase in the importance placed on maintaining data security. The digital supply chains of aerospace firms are protected by rigorous cybersecurity procedures that are implemented by the companies. The transition away from paper-based documentation results in fewer errors caused by human intervention and streamlines the operations. Traceability and auditability are both improved with the use of electronic records for things like orders, invoices, and certificates. The use of technologies that enable additive manufacturing enables the fabrication of components on demand, which reduces the need for large warehousing and helps to minimize lead times. The advent of 3D printing has made it possible to manufacture items in a manner that is both more flexible and decentralized, which has contributed to the creation of a supply chain that is both more agile and more responsive. The monitoring and assuring of compliance with regulatory standards is made easier with the use of digitization. Platforms are built with the capability to monitor changes in legislation and to automatically alter procedures in response to those changes.

AUGMENTED REALITY (AR) AND VIRTUAL REALITY (VR)

Augmented Reality (AR) and Virtual Reality (VR) have both found a place in the aviation business, where they are used for a variety of applications that improve training, maintenance, operations, and the overall customer experience. Training and Simulation, Cockpit Operations, Aircraft Design and Manufacturing, Maintenance and Inspection, Passenger Experience, Air Traffic Control (ATC) Training, Emergency Response Training, and Airports and Ground Operations are just few of the main areas where AR and VR are making an effect.

Virtual reality (VR) technology is integrated into flight simulators to create more lifelike training environments for pilots. Trainees can practice a variety of flying circumstances and emergency procedures in a safe setting thanks to this. In the field of aircraft maintenance instruction, augmented reality (AR) is used to superimpose digital information onto actual aircraft parts. Repairs and inspections can be made easier for technicians by providing them with step-by-step instructions and visual aids. AR is used in heads-up displays (HUDs), which project crucial flying information directly onto the windshield of an aircraft. This allows pilots to acquire crucial data without having to take their eyes off the outside environment, which significantly improves their situational awareness. Using augmented reality (AR) technology, navigational information and guidance signals can be projected onto the cockpit window to provide pilots with assistance in navigating and performing approach procedures.

During the design phase, virtual reality (VR) is deployed to provide engineers and designers with the ability to envision and interact with virtual prototypes of aircraft. The design process can be streamlined as a result, and possible problems can be found and resolved before actual prototypes are constructed. By superimposing digital instructions onto real-world components, augmented reality (AR) can facilitate the assembly process by directing workers through difficult assembly processes. During the process of airplane maintenance, AR applications give mechanics with real-time information about the various components of the aircraft. Digital overlays have the ability to call attention to particular regions that require attention and provide access to documentation that is relevant. AR makes it possible for specialists to offer remote support to maintenance teams by communicating in real time and exchanging visualizations, comments, and instructions with those teams. Virtual reality (VR) technology is utilized to provide passengers with fully immersive in-flight entertainment experiences. During flights, passengers can wear VR headsets to watch movies, participate in gaming or take virtual tours of the surrounding area. AR applications can provide passengers with pre-flight information, such as specifics about the aircraft, the flight path, and sites of interest, so elevating the quality of their travel experience as a whole.

Virtual reality (VR) is utilized in simulators for the purpose of providing air traffic controllers with a realistic setting in which to practice managing various air traffic scenarios. This makes it easier to make decisions and reduces the amount of time it takes to react in stressful situations. Virtual reality (VR) is used to train both aircraft and ground crews for emergency situations by simulating them. This provides the workers with the opportunity to train for and improve their response to a variety of emergency scenarios. AR navigation solutions provide ground personnel with real-time information regarding aircraft whereabouts, gate assignments, and other operational parameters, making it easier for them to navigate the airport environment. Virtual reality (VR) can be utilized for training in the procedures involved in the handling of baggage. This enables employees to practice loading and unloading scenarios in a simulated setting.

The application of augmented and virtual reality technology within the aviation sector results in improvements to passenger and employee safety, operational effectiveness, and the overall experience. These immersive technologies are anticipated to play a function that is of an increasingly crucial kind in molding the future of aviation as technology continues to improve.

CONCLUSION

The aerospace industry is constantly striving to improve its productivity, safety, and cost-effectiveness, which is driving the further incorporation of increasingly sophisticated forms of automation and robots into its operations. As technological development continues, anticipate additional inventions and uses in the aforementioned fields. The successful integration of data and connectivity within the aviation industry is vital for the creation of a more connected and collaborative environment, which will ultimately lead to improvements in operational efficiency, safety, and the overall passenger experience. The digitization of the supply chain in the aviation industry is a continuing process with the goals of increasing efficiency, lowering costs, and improving collaboration throughout the entirety of the aerospace supply chain. There is a good chance that the sector will witness additional innovations in digital solutions as technology continues to improve. These innovations will be made to answer the increasing issues of supply chain management.

REFERENCES

Aydn, S., & Kahraman, C. (2021). Aviation 4.0 Revolution. In *Intelligent and Fuzzy Techniques in Aviation 4.0: Theory and Applications* (pp. 3–19). Springer International Publishing.

ChenX.EderM. A.ShihavuddinA. S. M. (2023). A concept for human-cyber-physical systems of future wind turbines towards Industry 5.0. Authorea Preprints.

Elangovan, U. (2021). *Industry 5.0: The future of the industrial economy*. CRC Press. doi:10.1201/9781003190677

Ghobakhloo, M., Iranmanesh, M., Morales, M. E., Nilashi, M., & Amran, A. (2023). Actions and approaches for enabling Industry 5.0-driven sustainable industrial transformation: A strategy roadmap. *Corporate Social Responsibility and Environmental Management*, *30*(3), 1473–1494. doi:10.1002/csr.2431

Jafari, N., Azarian, M., & Yu, H. (2022). Moving from Industry 4.0 to Industry 5.0: What are the implications for smart logistics? *Logistics*, *6*(2), 26. doi:10.3390/logistics6020026

Jain, D. K., Li, Y., Er, M. J., Xin, Q., Gupta, D., & Shankar, K. (2021). Enabling unmanned aerial vehicle borne secure communication with classification framework for industry 5.0. *IEEE Transactions on Industrial Informatics*, *18*(8), 5477–5484. doi:10.1109/TII.2021.3125732

Javaid, M., & Haleem, A. (2020). Critical components of Industry 5.0 towards a successful adoption in the field of manufacturing. *Journal of Industrial Integration and Management*, *5*(03), 327–348. doi:10.1142/S2424862220500141

Kavya, J., Prasad, G., & Bharanidharan, N. (2022). Artificial Intelligence, Machine Learning, and Internet of Drones in Medical Applications. In *Bio-Inspired Algorithms and Devices for Treatment of Cognitive Diseases Using Future Technologies* (pp. 180–188). IGI Global. doi:10.4018/978-1-7998-9534-3.ch011

Massaro, A. (2021). *Electronics in Advanced Research Industries: Industry 4.0 to Industry 5.0 Advances*. John Wiley & Sons. doi:10.1002/9781119716907

Prasad, G. (2023). Internet of Unmanned Aerial Vehicle (IOU) in Industry 5.0. In Advanced Research and Real-World Applications of Industry 5.0 (pp. 178-188). IGI Global.

Prasad, G., Saravanan, K., Naveen, R., Nagaraj, V., Dinesh, R., & Anirudh, K. (2023). Influence of Machine Learning Technique in Unmanned Aerial Vehicle. In *Internet of Drones* (pp. 43–57). CRC Press. doi:10.1201/9781003252085-3

Sharma, R., & Arya, R. (2022). UAV based long range environment monitoring system with Industry 5.0 perspectives for smart city infrastructure. *Computers & Industrial Engineering*, *168*, 108066. doi:10.1016/j.cie.2022.108066

Thakur, P., & Sehgal, V. K. (2021). Emerging architecture for heterogeneous smart cyber-physical systems for industry 5.0. *Computers & Industrial Engineering*, *162*, 107750. doi:10.1016/j.cie.2021.107750

Zhang, C., Wang, Z., Zhou, G., Chang, F., Ma, D., Jing, Y., Cheng, W., Ding, K., & Zhao, D. (2023). Towards new-generation human-centric smart manufacturing in Industry 5.0: A systematic review. *Advanced Engineering Informatics*, *57*, 102121. doi:10.1016/j.aei.2023.102121

Chapter 7
Machine Learning-Based Environmental, Social, and Scientific Studies Using Satellite Images:
A Case Series

Rahul Ratnakumar
Manipal Institute of Technology, India & Manipal Academy of Higher Education, India

U. Vignesh
Vellore Institute of Technology, Chennai, India

ABSTRACT

Satellite images of the entire globe or any given location can be procured quickly with good resolution from any part of the world using thousands of artificial satellites revolving around the Earth. Using advanced AI/ML image processing algorithms, the acquired data can be analyzed to obtain various essential knowledge of any place at any time like the chemical composition of the environment, population density (social), etc. Another prominent field is military, defence, and warfare. Independent hyperspectral image cluster analysis of the world's heavily populated cities like Delhi, Shanghai, etc. has clearly shown the migration of population from rural to the urban. The RS-GIS technology combined with advanced machine learning algorithms predicts that only 50 more years of groundwater supply is left to be harvested. These case observations from different parts of the world show the power and scope of aero-oriented image processing using machine learning algorithms.

DOI: 10.4018/979-8-3693-1491-3.ch007

INTRODUCTION

Natural calamities like wild-fires, floods, earthquakes, volcanoes etc. and man-made disasters like nuclear, thermal or chemical accidents cannot be fully prevented (Demir et al., 2018). But with the sagacious design combined with the judicious use of technologies like real-time image segmentation (Ratnakumar et al., 2021) and allied architectures, quick actions can be executed to reduce their disastrous impact on humanity.

The Earth's vast canvas, once deciphered solely by trained eyes, is now being unravelled in intricate detail by the powerful duo of satellite imagery (Demir et al., 2018) and machine learning (Doshi, 2018). This ground-breaking marriage is revolutionizing our understanding of environmental disasters, from the destructive attacks of wildfires to the silent creep of deforestation. Traditional methods of monitoring wildfire and its spread rely on ground observation, often hampered by inaccessibility and danger. Satellite images, however, capture the inferno's fury from miles above, providing a real-time view of its path. Machine learning algorithms then step in, analysing the intricate patterns of heat, smoke, and vegetation change to predict the fire's trajectory with startling accuracy. This empowers firefighters to strategize their response, potentially saving lives and property. Beyond the immediate threat of wildfires, machine learning scours satellite images for the subtle scars of deforestation. By identifying changes in vegetation cover over time, algorithms (He, Kaiming et al., 2015) can pinpoint areas where trees have vanished, revealing the extent of illegal logging or the encroachment of agriculture into pristine forests. This information empowers policymakers to crack down on environmental crimes and implement sustainable forest management practices. The applications extend far beyond land-based disasters (Amit et al., 2017). Floods, for instance, leave their mark on the landscape in the form of altered water bodies and inundated vegetation. Machine learning algorithms trained on vast datasets of flood-affected areas can now analyse satellite images to detect floods in near real-time, providing crucial early warnings to vulnerable communities. This satellite-powered revolution in environmental disaster studies is still in its early stages, but the potential is immense. As machine learning algorithms (Rahman et al., 2016) become more sophisticated and data sets continue to grow, we can expect even greater accuracy and insights into the complex interplay between human actions and environmental change. From predicting the next wildfire to safeguarding our dwindling forests, the sky is no longer the limit – it's the launchpad for a new era of environmental intelligence (Voigt et al., 2016) By harnessing the power of machine learning (He et al., 2015) and satellite imagery, we are gaining a deeper understanding of our planet's vulnerabilities and the environmental forces that shape it. This knowledge is not just a technological marvel; it's a beacon of hope in the face of a changing world, empowering us to protect our planet and its inhabitants for generations to come.

When it comes to social studies- Satellite images can procure images of crowded cities 24x7 and compare them, perform complex operations on them and gain valuable social, human-behavioural insights in very less time. Machine learning algorithms can analyze these patterns, revealing insights into population density, economic activity, and even social well-being (Duarte et al., 2018). By tracking changes in light intensity over time, we can map urban growth, identify areas of poverty or deprivation, and even predict disease outbreaks.

Beyond the urban landscape, satellite images are playing a crucial role in scientific research. From tracking the migration patterns of endangered animals to monitoring the health of coral reefs, these images provide a bird's-eye view of complex ecological systems. Machine learning algorithms can identify subtle changes in vegetation cover, water quality, and animal behaviour, helping us understand the impact of climate change and human activities on the environment.

Real-time processing of satellite images requires state-of-the-art hardware (Ratnakumar et al., 2019). This includes high resolution multi-band image sensors, signal processing hardware architectures along with their renewable power sources embedded within the satellites, work in tandem to collect, analyse and evaluate the real-time data to retrieve valuable insights and crucial information about the situation at hand. Here the performance, size and efficiency are important along with the accuracy because of the severe space, resource and power constraints within these satellites (Ratnakumar et al., 2021). To overcome these constraints and achieve high performance from the ensuing hardware architecture the hardware designer must judiciously map the system functionalities into the embedded chip, so that the computation, size, complexity, power dissipation and durability of the ensuing hardware is optimized (Ratnakumar et al., 2019, January). Unlike power-hungry and number crunching conventional algorithms based on complex mathematical operations, evolutionary algorithms use no mathematical intricacies but simple yet powerful operations, which are adaptive, reconfigurable especially suited for hardware optimizations and supports fault tolerance. Reconfigurable computing hardware like Field Programmable Gate Arrays (FPGA) gives us an excellent platform to test and deploy the various architectural features of these algorithms in a hardware-aware perspective considering their speed, reconfigurability, fault tolerance, power consumption and area etc. Combining the fast reconfigurability of FPGA with the dynamic and evolutionary characteristics of the Nature Inspired Algorithms, opens the way to revolutionary opportunities like dynamic reprogramming, deploy and program concept, adaptive resource sharing, fault tolerance and quick troubleshooting.

CASE 1: THE MUIR GLACIER, AMAZON, YELLOWSTONE, TURKANA, AND URMIA

Figure 1. Muir glacier image taken by NASA

The Muir Glacier, one of the world's most beautiful places in Alaska, has attracted World's attention for more than twelve decades. A satellite image captured in 2004 shows how warmer temperatures have caused its shocking, 31-mile retreat (Fig. 1.). Similar eye-opening information from Satellite images reveal the critical state of the deep forests of Amazon facing heavy deforestation and alteration for industrial production. Machine learning algorithms predict the effects of deforestation that could cut the Amazon to 47% of its original size by 2030. Yellowstone, an American National park located in the Western United States, is a wonderland prone to wildfire. Satellite data informs us that it is steadily recovering after its 2 million- acres are washed away by forests-fire, which destroyed more than half of the park. The delta where the Omo river meets Africa's Lake Turkana was confined entirely within Ethiopia. However, recent satellite data reveals that it has grown so big it is now located mainly in Kenya. Multispectral images from MODIS and OLI of Iran's wonder lake Urmia which changed colour from green to red in a matter of weeks in the summer of 2016, invited serious investigation. The culprit- a combination of bacteria and algae that caused the change due to the rapid increase in temperature, the lake begins to evaporate, increasing its saltiness and reddish tinge. These observations from different parts of the world shows the power and scope of satellite image processing combined with Machine learning algorithms.

Figure 2. Muir glacier change over the years

Closely examining Muir Glacier over the past centuries, we can observe that in 1750, around the time the Little Ice Age reached its maximum, Glacier Bay in southeast Alaska was entirely filled with ice (Fig. 2.). When Captain George Vancouver sailed there in 1795, the glacial ice had pulled back to expose five miles of the bay—all part of a tidewater glacier's natural life cycle of advance and retreat. By the time naturalist John Muir explored and documented the region in 1879, the ice front had retreated 45 miles.

Today, the ice has retreated more than 60 miles (100 kilometers), and most vestiges of the original glacier now terminate on land. Just a handful of glaciers still flow into Glacier Bay and dispense icebergs—a type of glacier known as "tidewater glaciers." Only one—McBride Glacier—still flows into the bay's East Arm.

The past few decades of glacial retreat are visible in the images above, acquired on September 6, 1986, with Landsat 5, and September 17, 2019, with Landsat 8. Snow and ice are blue in these false-color images, which use different wavelengths to better differentiate areas of ice, rock, and vegetation. The images are centered on Glacier Bay's East Arm, which extends north past McBride Glacier and northwest into Muir Inlet.

Muir Inlet, Glacier Bay, and the surrounding land are part of Glacier Bay National Park. Around the time the park was established in 1980, most visitors viewed its glaciers from the vantage point of the water, where tidewater glaciers dropped huge slabs of ice into the bay. But the retreating glaciers, many of which now terminate on land, have changed how visitors experience the park.

CASE 2: MULTI-ANGLE IMAGER FOR AEROSOLS (MAIA)

Figure 3. Multi-Angle Imager for Aerosols (MAIA)

The estimates of the Global Burden of Disease, which studied 195 countries worldwide, reveals that in 2016 around 4.1 million of the world's population had died sooner than expected because of their exposure to airborne particles (Fig. 4.). However, in some countries, like India and China, the risk is much greater. Long-term exposures to air pollution are linked to lung cancer, heart disease, stroke and many other illnesses. Acute air pollution events have been linked to hospital admissions, asthma flare-ups and premature deaths. Breathing polluted air is linked to problems like high blood pressure during pregnancy and low birth weight in infants. In times of Volcanic eruptions, nuclear disaster or chemical gas leakage or explosions, the real-time understanding of the environmental chemical composition using real-time satellite data can save millions of lives (Sasaki et al., 2007). In the coming 2022 Jet Propulsion Laboratory, California Institute of Technology, NASA is planning to incorporating an instrument called Multi-Angle Imager for Aerosols (MAIA) into satellites, to investigate how the various types of air-pollution affect human health (Cao et al., 2018). These Instruments on space satellites can observe large areas of the Earth also can take up planned orbits targeting specific locations on Earth, like urban areas on different continents (Fig. 3). Unlike the digital camera used in our cell phone, which uses three filters to capture red, green, and blue wavelengths of light, the MAIA has 14 filters which can also detect ultraviolet, near-infrared, and shortwave-infrared bands which can provide information about the smallest and largest aerosol particles, even detect some organic matter as well. The MAIA instrument will collect data that can be used to calculate the physical properties of airborne particulate matter, like their shape, size, based on how they reflect or absorb sunlight. Using advanced image processing algorithms the acquired data can be analyzed to obtain the chemical composition of the environment (Duarte et al., 2018). This can be used for short term research, real-time disaster management as well as long term epidemiological studies (Fig. 4).

Figure 4. Mixtures of particulate matter constituents are most harmful to humans

CASE 3: METROPOLITAN CONVERSION

Figure 5. Satellite image of Delhi taken by NASA

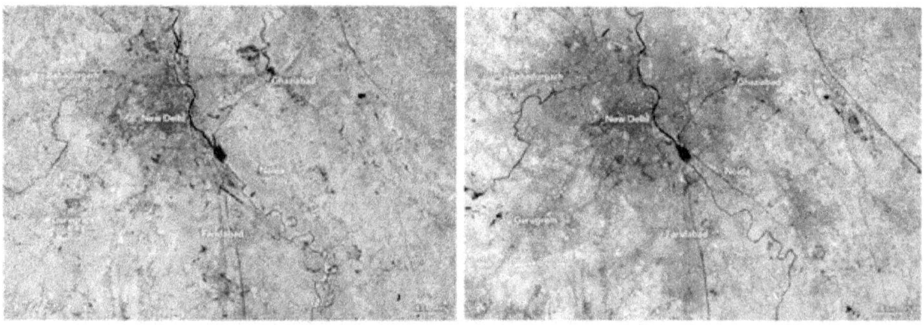

Satellite image analysis has revealed that in developing countries, an estimated 1 to 2 million hectares of cultivated lands are being converted every year to satisfy the demands of infrastructure, housing, industry etc. Built-up areas occupy up to 3% of the Earth's land surface. The research-based on satellite image segmentation analysis says- top in the list are the city of Dhaka Bangladesh and Ajmer city of Rajasthan in India an example of world's greatest urban sprawl. The global urban population has been increasing more rapidly than rural populations. Independent Hyperspectral image cluster analysis (Ignatiev et al., 2018) of the world's heavily populated cities like Delhi, India, with 19 Million residents and Shanghai, China, with 34 Million residents, have clearly shown the migration of population from rural to the urban. By 2030, Algorithms also predict a 30% percent conversion in the metropolitan area to built-up land. Approximately 1.2 million square kilometer of forests, woodlands and 5.6 million square kilometer of grassland, pasture lands have been transformed into other land types within the last three centuries. Such alterations in land-use cause significant impacts on Earth's climate.

The capital of India, New Delhi, has been experiencing one of the fastest urban expansions in the world. Vast areas of croplands and grasslands are being turned into streets, buildings, and parking lots, attracting an unprecedented amount of new residents. By 2050, the United Nations projects India will add 400 million urban dwellers, which would be the largest urban migration in the world for the thirty-two year period.

These images (Fig. 5) show the growth in the city of New Delhi and its adjacent areas—a territory collectively known as Delhi—from December 5, 1989, (left) to June 5, 2018 (right). These false-color images use a combination of visible and short-wave infrared light to make it easier to distinguish urban areas. The 1989 image was acquired by the Thematic Mapper on Landsat 5 (bands 7,5,3), and the 2018 image was acquired by Operational Land Imager on Landsat 8 (bands 7,6,4).

Most of the expansion in Delhi has occurred on the peripheries of New Delhi, as rural areas have become more urban. The geographic size of Delhi has almost doubled from 1991 to 2011, with the number of urban households doubling while the number of rural houses declined by half. Cities outside of Delhi—Bahadurgarh, Ghaziabad, Noida, Faridabad, and Gurugram—have also experienced urban growth over the past three decades, as shown in these images.

With a flourishing service economy, Delhi is a draw for migrants because it has one of India's highest per capita incomes. According to the latest census data, most people (and their families) move into the city for work. The Times of India reported that the nation's capital grew by nearly 1,000 people each day in 2016, of which 300 moved into the city. By 2028, New Delhi is expected to surpass Tokyo as the most populous city in the world.

The increased urbanization has had several consequences. One is that the temperatures of the urban areas are often hotter than surrounding vegetated areas. Manmade structures absorb the heat and then radiate that into the air at night, increasing the local temperature (the urban heat island effect). Research has shown that densely built parts of Delhi can be 7°C (45°F) to 9°C (48°F) warmer in the wintertime than undeveloped regions.

Additionally, sprawling cities can have several environmental consequences, such as increasing traffic congestion, greenhouse gas emissions, and air pollution. From 2005 to 2014, NASA scientists have observed an increase in air pollution in India due to the country's fast-growing economies and expanding industry.

India is one of many countries with fast-growing cities. By 2050, China is projected to add 250 million people in its urban areas, and Nigeria may add 190 million urban dwellers. In total, India, China and Nigeria are expected to account for 35 percent of the world's urban population growth between 2018 and 2050.

CASE 4: OIL FIELDS IN QAYYARAH

Figure 6. Qayyarah image taken by NASA

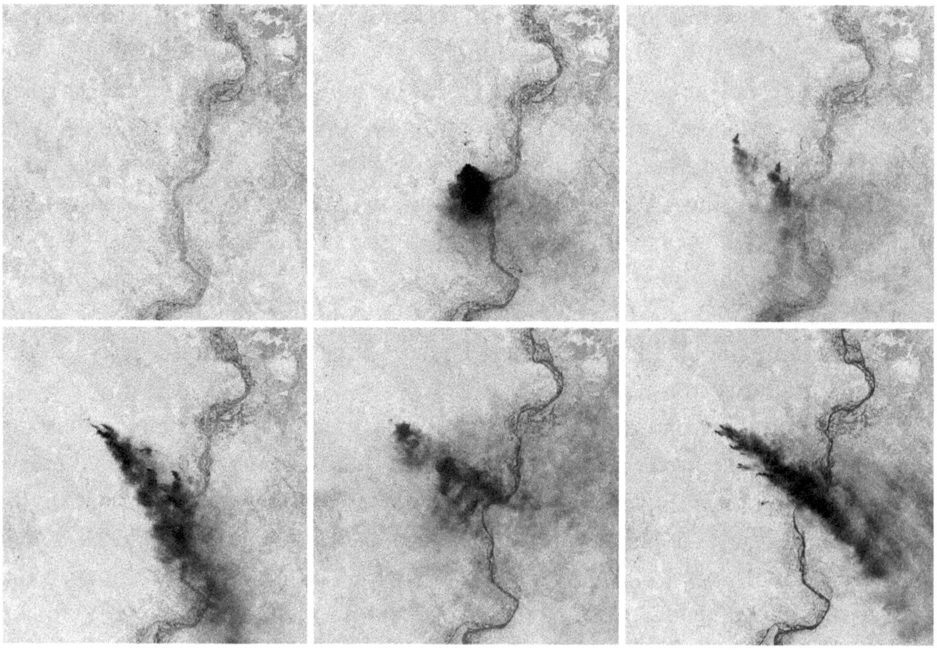

In May 2016, NASA's Landsat satellites first detected several devastating fires in the oil fields in Qayyarah, a small city on the banks of river Tigris, Iraq. The density of fires increased exponentially and took months to put out (Fig. 6.). However, unfortunately they were human-made atrocities. When the Islamic State of Iraq and the Levant (ISIL) on their retreat, bombed 700 oil wells, refineries and anything that supports the local economy. From July 2016 onward, satellites returned daily images showing clouds of thick, black smoke streaming from the Qayyarah. Scientists used LIDAR to quantify carbon-flux and deduce the carbon dioxide emitted. According to the estimates, these fires consumed roughly 4.6 million barrels of oil per day. The residents of the cities liberated from ISIL were forced into extremely polluted air, waterways, poisoned soil and filled with crude oil. The oil and gases poisoned Tigris river, which was the only source of drinking water and for watering crops. It took almost a year to control these fires. The fire depended on many factors like the type of fuel, humidity, wind, and fuel density also affect the smoke's color. The after-effects were disastrous, including a "nuclear winter"- extreme cooling effect at Earth's surface (Sasaki et al., 2007). The airborne smoke particles blocked Sun by reflecting back the sunlight towards space, driving down temperatures on the ground. The particles in the oil smoke would rise into the stratosphere and stay aloft for weeks or months over the Northern Hemisphere, which could be estimated using real-time satellite image analysis (Fujita et al., 2017) and machine learning (Duarte et al., 2018), saving thousands of lives in the process.

Several oil fires that have burned in northern Iraq for months have been extinguished. NASA satellites first detected fires at one or two wells near the towns of Qayyarah and Shargat in May 2016. Small fires continued to spring up intermittently throughout June. In July 2016, the number and volume of fires increased dramatically.

From July 2016 onward, satellites returned daily images showing clouds of thick, black smoke streaming from the Qayyarah oil field. For a period in October, black smoke from oil fires even mixed with a noxious white cloud of sulfur dioxide from a separate blaze at a factory.

By April 2017, Iraqi firefighters had put out the fires. NASA satellites last detected heat and smoke from the oil field on March 27, 2017. Iraq's oil ministry made a formal announcement about the end of the fires on April 4, 2017. Of the 50 oil wells in the area, militants lit 18 of them on fire.

The Operational Land Imager (OLI) on Landsat 8 captured these false-color images on January 24, 2017, and March 29, 2017. The images combine shortwave-infrared, near-infrared, and green light (OLI bands 6-5-3) to show active fires (bright red) and smoke (black). The vegetation (green) in the March 29 image is the product of spring plant growth.

CASE 5: DESERT OF WADI-AL-SARHAN

Figure 7. Desert of Wadi-al-Sarhan image taken by NASA

Converting deserts into massive agricultural hubs is not impossible with the help of current technology. By tapping the water resources beneath the sands of the Saudi Arabian desert, engineers turned the same desert of Wadi-al-Sarhan into an oasis in around two decades (Fig. 7.). By searching and finding water under the desert by Hyperspectral analysis of specific spectral bands which can match the reflective coefficient of water. Remote sensing (RS) and the geographical information system (GIS) techniques have been proven to be useful tools to depict spatial and temporal changes in land cover in these oases. The same technology, combined with advanced machine learning algorithms predicts that only 50 more years of groundwater supply is left to be harvested. So it is now a race against time.

Northern Saudi Arabia hosts some of the most extensive sand and gravel deserts in the world, but modern agriculture has changed the face of some of them. This photograph from astronauts on the International Space Station presents an almost surreal view of abundant green fields in the midst of a barren desert.

As recently as 1986, there was little to no agricultural activity in the Wadi As-Sirhan Basin. But over the past 26 years, agricultural fields have been steadily developed, largely as a result of the investment of oil industry revenues by the Saudi government. Crops grown in the area include fruits, vegetables, and wheat.

The fields are irrigated by water pumped from underground aquifers. That water is distributed in rotation about a center point within a circular field—a technique known as center-pivot agriculture. The approach affords certain benefits compared to traditional surface irrigation, such as better control of water and fertilizer use. This so-called "precision agriculture" is particularly important in regions subject to high water loss due to evaporation. By better controlling the amount and timing of water application, evaporative losses can be minimized.

For a sense of scale, the agricultural fields in active use (dark green) or fallow (brown to tan), are approximately one kilometer in diameter. While much of the Wadi As-Sirhan Basin shown here is sandy (light tan to brown surfaces) and relatively flat, low hills and rocky outcrops (dark gray) of underlaying sedimentary rocks are visible at image left and right.

Courtesy (Fig.7): Astronaut photograph ISS030-E-90918 was acquired on February 21, 2012, with a Nikon D2Xs digital camera using a 65 mm lens, and is provided by the ISS Crew Earth Observations experiment and Image Science & Analysis Laboratory, Johnson Space Center. The image was taken by the Expedition 30 crew.

CONCLUSION

Machine learning's integration into environmental, social, and scientific studies through satellite image analysis has propelled multifaceted advancements in our understanding of Earth's dynamics. By employing sophisticated algorithms, these applications decipher satellite data to uncover a spectrum of insights crucial for diverse fields. Environmental studies benefit from ML models that detect changes in ecosystems, aiding in biodiversity conservation and monitoring climate shifts. In social studies, these technologies reveal socioeconomic trends, urban development, and population dynamics, informing policy decisions and resource allocation. Additionally, in scientific research, satellite-based ML analyses contribute to understanding geological phenomena, weather patterns, and natural disasters. The convergence of machine learning and satellite imagery amplifies our capacity to address pressing global challenges by providing comprehensive, data-driven perspectives across environmental, social, and scientific domains. These multiple case observations from different parts of the world discussed in this chapter shows the power and scope of satellite image processing using Machine learning algorithms.

REFERENCES

Amit, S. N. K. B., & Aoki, Y. (2017), Disaster detection from aerial imagery with convolutional neural network. *Knowledge Creation and Intelligent Computing (IES-KCIC), 2017 International Electronics Symposium (IEEE),* 239–245 10.1109/KCIC.2017.8228593

Cao, Q. D., & Choe, Y. (2018), Deep Learning Based Damage Detection on Post Hurricane Satellite Imagery. *arXiv preprint arXiv:1807.01688.*

Demir, Koperski, Lindenbaum, Pang, Huang, Basu, Hughes, Tuia, & Raskar. (2018). *Deepglobe: A challenge to parse the earth through satellite images.* ArXiv e-prints.

Doshi, J. (2018), Residual Inception Skip Network for Binary Segmentation. *Proceedings of the IEEE Conference on Computer Vision and Pattern Recognition Workshops,* 216–219. 10.1109/CVPRW.2018.00037

Duarte, D., Nex, F., Kerle, N., & Vosselman, G. (2018). Satellite Image Classification of building damages using airborne and satellite image samples in a deep learning approach. *ISPRS Annals of the Photogrammetry, Remote Sensing and Spatial Information Sciences, 4*(2), 89–96. doi:10.5194/isprs-annals-IV-2-89-2018

Fujita, A., Sakurada, K., Imaizumi, T., Ito, R., Hikosaka, S., & Nakamura, R. (2017). Damage detection from aerial images via convolutional neural networks. *IAPR International Conference on Machine Vision Applications, Nagoya, Japan,* 8–12. 10.23919/MVA.2017.7986759

He, K., Zhang, X., Ren, S., & Sun, J. (2015). Delving deep into rectifiers: Surpassing human-level performance on imagenet classification. *Proceedings of the IEEE international conference on computer vision,* 1026–1034. 10.1109/ICCV.2015.123

Ignatiev, V., Trekin, A., Lobachev, V., Potapov, G., & Burnaev, E. (2018). Targeted change detection in remote sensing images. *arXiv preprint arXiv:1803.05482.*

Rahman, M. A., & Wang, Y. (2016). Optimizing intersection-over-union in deep neural networks for image segmentation. *International Symposium on Visual Computing.* 234–244. 10.1007/978-3-319-50835-1_22

Ratnakumar, R., & Nanda, S. J. (2019). A low complexity hardware architecture of K-means algorithm for real-time satellite image segmentation. *Multimedia Tools and Applications, 78*(9), 11949–11981. doi:10.1007/s11042-018-6726-6

Ratnakumar, R., & Nanda, S. J. (2021). A high speed roller dung beetles clustering algorithm and its architecture for real-time image segmentation. *Applied Intelligence, 51*(7), 4682–4713. doi:10.1007/s10489-020-02067-7

Ratnakumar, R., & Nanda, S. J. (2021, January). An improved genetic clustering architecture for real-time satellite image segmentation. In *2021 International Conference on Advances in Technology, Management & Education (ICATME)* (pp. 123-128). IEEE. 10.1109/ICATME50232.2021.9732768

Sasaki, Y. (2007). The truth of the F-measure. *Teach Tutor mater, 1*(5), 1–5.

Voigt, S., Giulio-Tonolo, F., Lyons, J., Kucera, J., Jones, B., Schneiderhan, T., Platzeck, G., Kaku, K., Hazarika, M. K., Czaran, L., Li, S., Pedersen, W., James, G. K., Proy, C., Muthike, D. M., Bequignon, J., & Guha-Sapir, D. (2016). Global trends in satellite-based emergency mapping. *Science, 353*(6296), 247–252. doi:10.1126/science.aad8728 PMID:27418503

Chapter 8
Machine Learning for Aerospace Object Categorization

Pavan Vignesh
https://orcid.org/0009-0006-0884-295X
Vellore Institute of Technology, Chennai, India

R. Maheswari
Symbiosis Skills and Professional University, India

P. Vijaya
Modern College of Business and Science, Oman

U. Vignesh
Vellore Institute of Technology, Chennai, India

ABSTRACT

Star classification plays a crucial role in understanding the vast and complex universe. Traditional methods of classifying stars are often labour-intensive and time-consuming. The advent of machine learning techniques has opened up new possibilities for automating and improving the accuracy of star classification. This chapter presents an overview of an interstellar star classification system based on machine learning. In this study, the authors propose a novel approach to classify stars using machine-learning algorithms. The dataset comprises a comprehensive collection of stellar data, including spectral characteristics, luminosity, temperature, and other relevant features. The goal is to develop a robust and accurate classification model that can categorize stars into various spectral classes, such as O, B, A, F, G, K, and M, along with their luminosity classes. This study demonstrates how machine learning could revolutionize the classification of aircraft objects, opening the door to improvements in aeronautical engineering.

DOI: 10.4018/979-8-3693-1491-3.ch008

1. INTRODUCTION

Star classification plays a crucial role in understanding the vast and complex universe. Traditional methods of classifying stars are often labour-intensive and time-consuming. The advent of machine learning techniques has opened up new possibilities for automating and improving the accuracy of star classification. This abstract presents an overview of an interstellar star classification system based on machine learning. In this study, we propose a novel approach to classify stars using machine-learning algorithms. The dataset comprises a comprehensive collection of stellar data, including spectral characteristics, luminosity, temperature, and other relevant features. Our goal is to develop a robust and accurate classification model that can categorize stars into various spectral classes, such as O, B, A, F, G, K, and M, along with their luminosity classes (Poduval et al., 2023).

Traditional stellar classification methods rely on manual analysis of stellar spectra, which is a time-consuming and laborious process. With the advent of large-scale astronomical surveys and the increasing volume of observational data, there is a growing need for automated and efficient classification techniques (Azari et al., 2021).

Deep Learning for Stellar Classification

Deep learning, a subset of machine learning, has emerged as a powerful tool for analyzing complex data patterns, including stellar spectra (Abubakar et al., 2019). Deep learning algorithms can learn to identify subtle features and relationships in data, making them well-suited for automated stellar classification tasks.

Benefits of Deep Learning-Based Analysis

Deep learning-based approaches offer several advantages over traditional methods for stellar classification:

1. Accuracy: Deep learning models can achieve high accuracy in classifying stars, outperforming traditional methods in many cases.
2. Scalability: Deep learning models can be trained on large datasets, making them well-suited for analyzing large-scale astronomical surveys.
3. Efficiency: Deep learning models can automate the classification process, significantly reducing the time and effort required compared to manual methods.
4. Feature Extraction: Deep learning models can learn to extract relevant features from stellar spectra without the need for prior knowledge or manual feature engineering.

Applications of Deep Learning-Based Stellar Classification

Deep learning-based stellar classification has a wide range of applications in astronomy, including:

1. Identifying and characterizing stars in large astronomical surveys.
2. Understanding stellar evolution and the formation of galaxies.
3. Studying the composition and properties of stars in different environments.
4. Identifying potential exoplanet host stars.

Future Directions

Deep learning-based stellar classification is a rapidly developing field, and there is ongoing research to improve the accuracy, efficiency, and applicability of these techniques. Future directions include:

Developing more sophisticated deep learning architectures specifically tailored for stellar classification tasks.

Incorporating additional data sources, such as multi-band photometry and imaging data, to improve classification performance.

Exploring transfer learning approaches to leverage existing deep learning models for stellar classification.

Developing Explainable AI (XAI) techniques to understand the decision-making processes of deep learning models for stellar classification.

Deep learning-based analysis has the potential to revolutionize stellar classification, enabling astronomers to make more efficient and accurate discoveries about the universe.

2. RELATED WORKS

Astronomy and astrophysics have witnessed significant advancements in recent decades, and the exploration of stellar classification has been a cornerstone in understanding the vast celestial landscape. Various studies and datasets have contributed to the field, with a focus on unraveling the mysteries of stars and their diverse properties (Ting & Ranganath, 2019).

One notable dataset that has paved the way for stellar classification research is the Sloan Digital Sky Survey (SDSS). SDSS has been instrumental in providing vast amounts of astronomical data, including spectra and photometric measurements for millions of celestial objects, including stars. The earlier releases of SDSS, such as SDSS-III, played a crucial role in mapping the Milky Way and identifying stars based on their spectral signatures. This dataset laid the foundation for understanding stellar populations and their distribution within our galaxy (Breiman, 2001).

In parallel, efforts have been made to create specialized datasets dedicated explicitly to stellar classification. The Geneva-Copenhagen Survey (GCS) is one such example. It is a compilation of stellar properties for a large sample of nearby stars, providing detailed information on stellar ages, metallicities, and kinematics.

This dataset has been pivotal in advancing our knowledge of the Milky Way's stellar population, enabling more precise classification and analysis of different stellar types (Aihara et al., 2011).

Advancements in technology and observational techniques have led to the creation of datasets with enhanced capabilities. The Gaia mission, launched by the European Space Agency, has been a game-changer in this regard. Gaia has been meticulously measuring the positions, distances, and motions of over a billion stars, providing an unprecedented 3D map of our galaxy. This dataset has not only revolutionized our understanding of stellar dynamics but has also opened new avenues for precise stellar classification based on a wealth of observational parameters (Lépine & Shara, 2005).

Machine learning techniques have increasingly become integral to stellar classification research, with datasets serving as the training ground for sophisticated models. The introduction of convolutional neural networks (CNNs) and recurrent neural networks (RNNs) has allowed for more intricate analyses of stellar spectra and photometric data. The LAMOST Spectroscopic Survey of the Milky Way (LAMOST) has contributed significantly in this domain. LAMOST has provided a vast dataset of stellar spectra, allowing researchers to apply machine learning algorithms to classify stars based on their spectral features. This dataset has been crucial for developing and refining automated methods for stellar classification (Perryman, 2001).

Furthermore, efforts have been directed towards creating datasets that encompass a broader range of stellar types, including variable stars and exotic objects. The OGLE (Optical Gravitational Lensing Experiment) survey, for instance, has focused on monitoring variable stars, providing a dataset rich in time-series data. This has implications not only for understanding stellar pulsations but also for identifying and classifying variable stars with applications in cosmology.

The advent of large-scale sky surveys, such as the Pan-STARRS (Panoramic Survey Telescope and Rapid Response System), has also significantly contributed to stellar classification efforts. Pan-STARRS has provided multi-band photometric data for billions of objects, offering a comprehensive view of the sky. This dataset has been leveraged for classifying stars based on their colors, luminosities, and positions in the Hertzsprung-Russell diagram, advancing our understanding of stellar evolution.

The utilization of spectroscopic surveys has been a recurring theme in stellar classification research. The RAVE (RAdial Velocity Experiment) survey, for example, has focused on measuring radial velocities for stars, providing crucial information for understanding the dynamics of the Milky Way. This dataset has been employed in studies aiming to classify stars based on their kinematic properties, adding a new dimension to stellar taxonomy (de Bruijne, 2012).

In recent years, the exploration of exoplanetary systems has become a focal point in astronomy. Datasets like the Kepler Space Telescope mission data have not only led to the discovery of numerous exoplanets but have also presented opportunities for studying host stars. Stellar properties derived from these datasets have contributed to refining our understanding of stellar evolution and have facilitated the identification of stars hosting potentially habitable exoplanets (Brown et al., 2018).

In conclusion, the field of stellar classification has evolved significantly, propelled by the availability of diverse and expansive datasets. From pioneering surveys like SDSS to revolutionary missions like Gaia, these datasets have been the bedrock of astronomical research. Combining traditional observational methods with state-of-the-art machine learning techniques, researchers continue to push the boundaries of stellar classification, unlocking the secrets of the universe one star at a time. The ongoing and future astronomical surveys promise to provide even more extensive and detailed datasets, offering new avenues for exploration and discovery in the captivating realm of astrophysics.

2.1 Machine Learning-Based Analysis

The proposed methodology involves several key steps:

Data Collection and Pre-processing: A diverse and well-curated dataset of stars is collected, spanning different spectral types and luminosity classes. The data is pre-processed to handle missing values, outliers, and normalization.

Feature Extraction and Selection: Relevant features are extracted from the stellar dataset, taking into account physical attributes and spectroscopic data. Feature selection techniques are employed to choose the most informative attributes.

Model Selection and Training: Several machine learning methods, like random forests, decision trees, support vector machines, and neural networks, get evaluated for their performance in the classification of stars. The models are built using cross-validation techniques after being trained on a labelled set of data.

Performance Evaluation: The trained models are rigorously evaluated using metrics such as accuracy, precision, recall, and F1-score. Confusion matrices are analyzed to understand the model's strengths and weaknesses in classifying different types of stars.

Interpretability and Visualization: To enhance the interpretability of the classification process, techniques like feature importance analysis and visualization of decision boundaries are employed.

Comparative Analysis: The performance of the proposed machine learning-based star classification system is compared with traditional methods to demonstrate its superiority in terms of accuracy and efficiency.

Deployment and Future Directions: The developed model can be integrated into astronomical software tools and databases for real-time star classification. Future work may involve expanding the dataset, exploring transfer learning, and incorporating additional features for even finer classification granularity. In conclusion, the interstellar star classification system presented in this abstract leverages the power of machine learning to accurately categorize stars based on their spectral characteristics and luminosity classes. The results indicate the potential for significant advancements in the field of astronomy and astrophysics, paving the way for deeper insights into the nature of the universe.

1. A machine learning-driven technique for enhancing aircraft item categorization is provided. In the field of aeronautical science, it is important to correctly categorize aerospace objects. The research suggests using machine learning methods to simplify and automate this process. The process involves gathering data, identifying features, and training models using a variety of algorithms, including convolutional neural networks and support vector machines (LeCun et al., 2015). Precision, recall, and F1-score are evaluation metrics which demonstrate how well the model performs. The analysis of feature importance improves the model's interpretability. The outcomes show that traditional methods don't perform as well in terms of accuracy and efficiency (He et al., 2016). This study demonstrates how machine learning could revolutionize the classification of aircraft objects, opening the door to improvements in aeronautical engineering (Kingma & Ba, 2014).

2.2 Deep Learning Based Analysis

Deep learning has emerged as a powerful tool in the field of astrophysics, particularly in the analysis and classification of stellar datasets. The application of deep learning techniques to Stellar Classification Datasets (SCDs) enhances our ability to discern intricate patterns in stellar spectra, enabling more accurate and efficient classification (Simonyan & Zisserman, 2014).

Here, we delve into various deep learning approaches employed in stellar classification research:

Convolutional Neural Networks (CNNs):

1. Spectral Feature Extraction:

Convolutional Neural Networks have proven effective in automatically learning hierarchical features from stellar spectra. By applying convolutional filters, CNNs can identify complex spectral patterns indicative of different stellar types. This approach has been instrumental in automating the classification process, reducing the reliance on manual feature engineering.

2. Transfer Learning:

Transfer learning, leveraging pre-trained CNN models on general image datasets, has been adapted for stellar classification. The learned features from unrelated datasets can be fine-tuned on stellar spectra, optimizing model performance even with limited labeled data (Srivastava et al., 2014).

Recurrent Neural Networks (RNNs):

1. Time-Series Analysis:

For datasets containing time-series information, such as variable stars, Recurrent Neural Networks excel in capturing temporal dependencies. RNNs can analyze variations in stellar brightness over time, aiding in the classification of variable stars and understanding their pulsation characteristics.

2. Long Short-Term Memory (LSTM) Networks:

LSTMs, a type of RNN, are particularly useful for handling sequences of data. In the context of stellar classification, LSTMs can effectively model the temporal evolution of spectral features, contributing to improved accuracy in identifying different stellar classes.

Hybrid Architectures:

1. Combining CNNs and RNNs:

Hybrid architectures that integrate both CNNs and RNNs have demonstrated success in capturing both spectral and temporal information. This is particularly beneficial when dealing with datasets containing multi-dimensional information, such as spectroscopic time-series data.

Generative Models:

1. Variational Autoencoders (VAEs):

VAEs can be employed for unsupervised representation learning. By training on stellar spectra, VAEs can generate latent space representations that capture essential features, aiding in clustering or classification tasks (Chollet, 2015).

Challenges and Considerations:

1. Limited Labeled Data:

Stellar classification datasets often face challenges of limited labeled data for specific classes. Techniques like data augmentation and transfer learning become crucial in addressing this issue.

2. Interpretability:

Interpreting the decisions of deep learning models in astrophysics is an ongoing challenge. Understanding how and why a model classifies a star into a particular category is essential for building trust in automated systems.

3. Generalization to New Phenomena:

Ensuring that deep learning models generalize well to new and unseen phenomena is crucial. Continuous efforts are made to enhance the robustness of models in handling diverse and evolving datasets.

Future Directions:

1. Explainable AI: Integrating explainable AI techniques to enhance the interpretability of deep learning models will be pivotal in gaining insights into the features driving stellar classifications.
2. Active Learning:

Implementing active learning strategies to intelligently select and label the most informative data points can help in optimizing model performance with limited labeled data.

3. Incorporating Multi-Wavelength Data:

Fusion of multi-wavelength data, combining information from different parts of the electromagnetic spectrum, is a promising avenue for enhancing the capabilities of deep learning models in stellar classification.

In conclusion, the application of deep learning in stellar classification datasets represents a paradigm shift in the way astronomers analyze and interpret celestial data. The adaptability of these models to learn intricate patterns, coupled with advancements in interpretability and generalization, continues to push the boundaries of our understanding of stars and their diverse properties. As datasets expand in scale and complexity, the synergy between deep learning methodologies and astrophysical research is poised to unlock new frontiers in the exploration of the cosmos.

2.3 Block Diagram

Figure 1. Block diagram

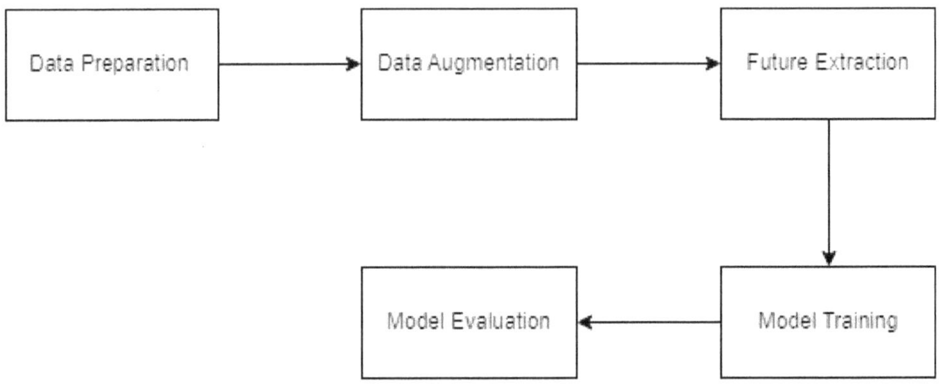

2.4 Architecture

Figure 2. Architecture

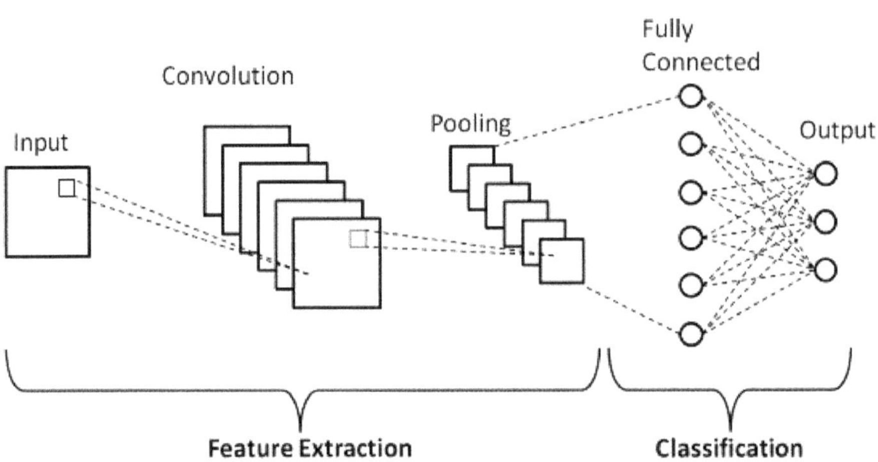

2.4.1 Preprocessing

- Data Normalization: Normalize the stellar spectra to a consistent scale to ensure that all features have equal importance (Fig. 2.).

- Noise Removal: Apply noise reduction techniques to remove unwanted noise from the spectra.
- Feature Scaling: Scale the features to a suitable range for the deep learning model.

2.4.2 Convolutional Neural Network (CNN)

- Convolutional Layers: Extract local features from the stellar spectra using convolutional filters.
- Pooling Layers: Reduce the dimensionality of the feature maps while retaining essential information.
- Activation Functions: Introduce non-linearity into the network using activation functions like ReLU or Leaky ReLU.
- Batch Normalization: Normalize the activations of each layer to stabilize the training process (Gaia Collaboration et al., 2018).

2.4.3 Fully Connected (FC) Layers

- Hidden Layers: Learn high-level features and relationships between the extracted features.
- Dropout Regularization: Apply dropout regularization to prevent overfitting and improve generalization.
- Output Layer: Map the high-level features to the final classification probabilities for each stellar class.

2.4.4 Training and Validation

- Training Process: Optimize the model parameters using a gradient-based optimization algorithm like Adam or SGD.
- Loss Function: Minimize a loss function, such as cross-entropy, to measure the discrepancy between the predicted and true labels.
- Validation Set: Evaluate the model's performance on a separate validation set to prevent overfitting.

2.4.5 Interpretation and Evaluation

- Visualize Features: Explore the learned features to understand how the model distinguishes between stellar classes.
- Confusion Matrix: Analyze the classification performance by examining the confusion matrix.
- Metrics: Evaluate the model's accuracy, precision, recall, and F1-score using appropriate metrics.

This architecture provides a general framework for deep learning-based stellar classification. Specific details, such as the number of layers, filter sizes, and activation functions, may vary depending on the size and complexity of the dataset.

2.5 Results

Figure 3. Properties of star

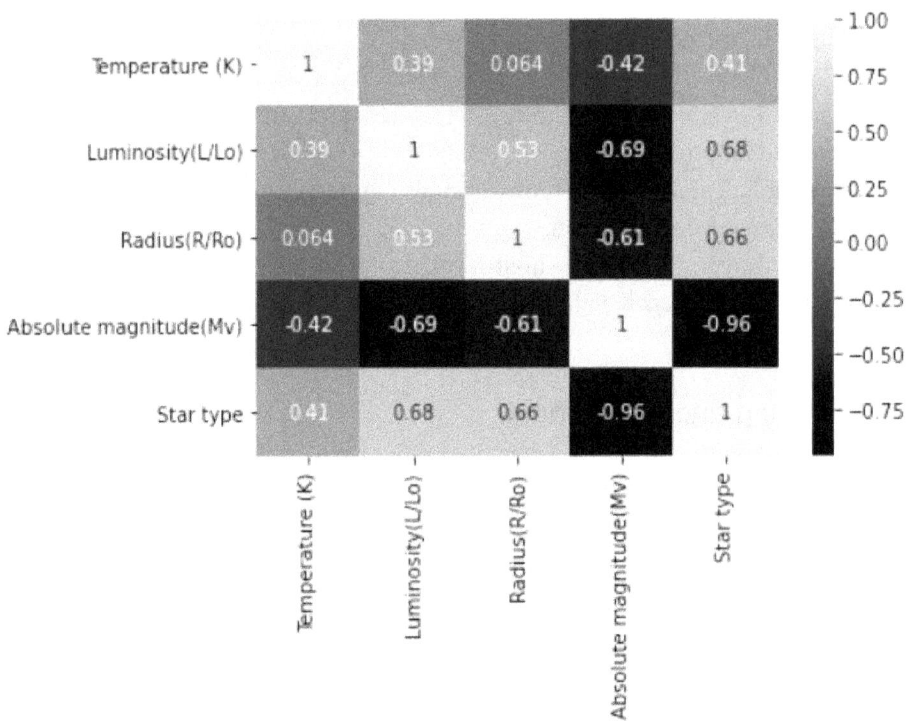

The features we have here are a few properties of a star (Fig. 3.):

- Absolute Temperature (in K) Temperature of the Star in Kelvin
- Relative Luminosity (L/Lo) Relative Luminosity is the ratio of brightness (Luminosity) of the star with the brightness of the Sun.
- Relative Radius (R/Ro) Relative Radius is again the ratio of the radius of a star compared to the radius of the Sun.
- Absolute Magnitude (Mv) Absolute magnitude (M) is a measure of the luminosity of a celestial object, on an inverse logarithmic astronomical magnitude scale.
- Star Color (white,Red,Blue,Yellow,yellow-orange etc) Self explainatory. It is the colour of Star as it appears in the sky.
- Spectral Class (O,B,A,F,G,K,,M) It is basically a class in which a star falls based on most of the above features.
- Star Type (Red Dwarf, Brown Dwarf, White Dwarf, Main Sequence, SuperGiants, HyperGiants) Type of star. We'll be trying to predict this value using the features

The first plot (Fig. 4.) illustrates the training and validation loss, and the second plot (Fig. 5.) illustrates the training and validation accuracy. These visualizations are useful for assessing the performance and training progress of a machine learning model during different epochs of training. The legend function is used to label the lines in the plot, and the savefig function is used to save the plots as image files.

Figure 4. Training loss vs. validation loss

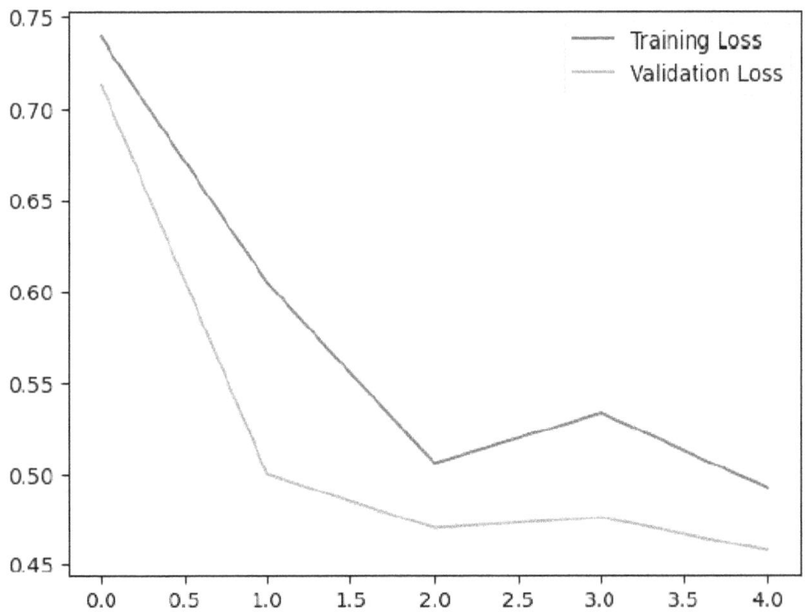

Figure 5. Training accuracy vs. validation accuracy

The image Fig. 6. creates a horizontal count plot using Seaborn to visualize the distribution of the "Spectral Class" variable in the dataset. The figure size is set to be 12x6 inches, and the color palette is chosen to be 'husl'. The resulting plot will show the counts of each unique value in the "Spectral Class" column, with bars arranged along the y-axis. This type of visualization is helpful for understanding the distribution of categorical variables in a dataset.

Figure 6. Count for each spectral class

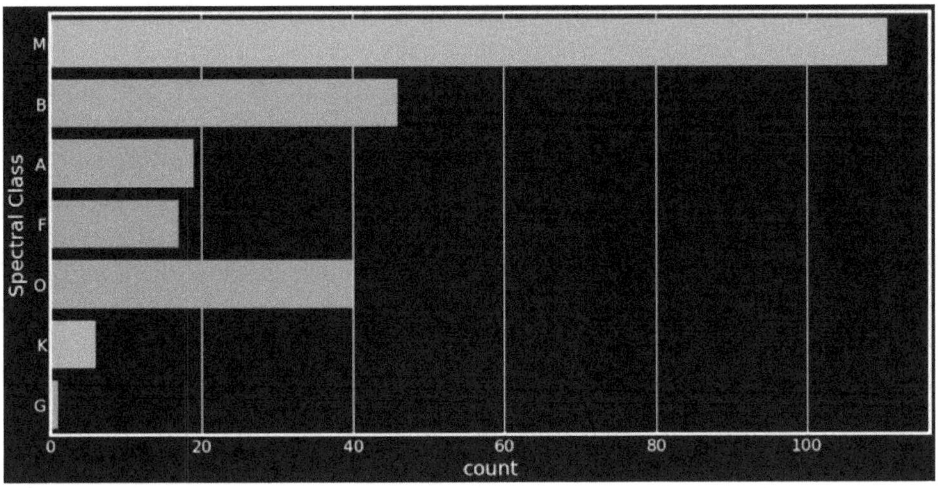

This image Fig 6 creates a horizontal count plot using Seaborn to visualize the distribution of the "Star type" variable in the dataset. The figure size is set to be 12x6 inches, and the color palette is chosen to be 'husl'. The resulting plot will show the counts of each unique value in the "Star type" column, with bars arranged along the y-axis. This type of visualization is useful for understanding the distribution of different star types in the dataset.

Figure 7. Values for each star type

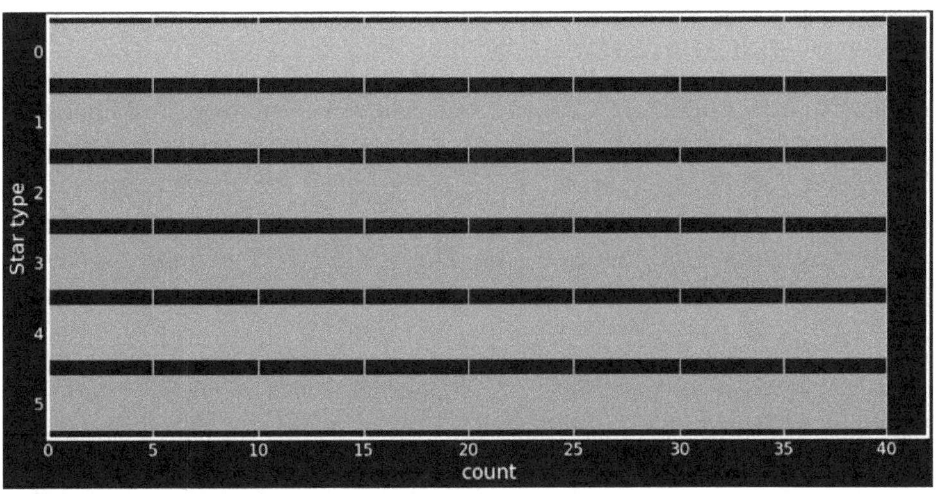

3. CONCLUSION

Satellite Image Classification represents a transformative application of machine learning in extracting meaningful information from vast sets of satellite imagery. Through the utilization of advanced algorithms, it enables the automated identification and categorization of diverse features such as land cover, infrastructure, and environmental changes. This technology holds immense potential for applications in agriculture, urban planning, disaster management, and environmental monitoring. As it continues to evolve, Satellite Image Classification stands at the forefront of harnessing the power of artificial intelligence to unlock actionable insights from the wealth of visual data captured by satellites, fostering advancements across various domains.The utilization of star data for predicting star types signifies a significant stride in astrophysical understanding. Applying machine learning models to star datasets enables the discernment of distinctive stellar characteristics, fostering insights into the vast celestial landscape. By leveraging advanced algorithms, this predictive approach enhances our ability to classify stars based on their unique attributes, contributing to a deeper comprehension of stellar evolution, composition, and behavior. The synergy between astronomy and machine learning, as demonstrated in star type prediction, opens avenues for uncovering celestial mysteries and enriching our knowledge of the diverse and captivating entities that populate the cosmic tapestry.

REFERENCES

Abubakar, A., Kolo, K., & Zhang, Y. (2019). Machine Learning Approaches in Astrophysics: A Review. *Frontiers in Astronomy and Space Sciences*, *6*, 53. doi:10.3389/fspas.2019.00053

Aihara, H., Allende Prieto, C., An, D., Anderson, S. F., Aubourg, É., Balbinot, E., Beers, T. C., Berlind, A. A., Bickerton, S. J., Bizyaev, D., Blanton, M. R., Bochanski, J. J., Bolton, A. S., Bovy, J., Brandt, W. N., Brinkmann, J., Brown, P. J., Brownstein, J. R., Busca, N. G., ... Zehavi, I. (2011). The Eighth Data Release of the Sloan Digital Sky Survey: First Data from SDSS-III. *The Astrophysical Journal. Supplement Series*, *193*(2), 29. doi:10.1088/0067-0049/193/2/29

Azari, A., Biersteker, J. B., Dewey, R. M., Doran, G., Forsberg, E. J., Harris, C. D. K., Kerner, H. R., Skinner, K. A., Smith, A. W., Amini, R., Cambioni, S., Poian, V. D., Garton, T. M., Himes, M. D., Millholland, S., & Ruhunusiri, S. (2021). Integrating Machine Learning for Planetary Science: Perspectives for the Next Decade. *Bulletin of the AAS*, *53*(4). Advance online publication. doi:10.3847/25c2cfeb.aa328727

Breiman, L. (2001). Random forests. *Machine Learning*, *45*(1), 5–32. doi:10.1023/A:1010933404324

Brown, A. G. A., Vallenari, A., Prusti, T., de Bruijne, J. H. J., Babusiaux, C., Bailer-Jones, C. A. L., Biermann, M., Evans, D. W., Eyer, L., Jansen, F., Jordi, C., Klioner, S. A., Lammers, U., Lindegren, L., Luri, X., Mignard, F., Panem, C., Pourbaix, D., Randich, S., ... Zwitter, T. (2018). Gaia Data Release 2: Summary of the contents and survey properties. *Astronomy & Astrophysics*, *616*, A1. doi:10.1051/0004-6361/201833051

Brown, A. G. A., Vallenari, A., Prusti, T., de Bruijne, J. H. J., Babusiaux, C., Bailer-Jones, C. A. L., Biermann, M., Evans, D. W., Eyer, L., Jansen, F., Jordi, C., Klioner, S. A., Lammers, U., Lindegren, L., Luri, X., Mignard, F., Panem, C., Pourbaix, D., Randich, S., ... Zwitter, T.Gaia Collaboration. (2018). Gaia Data Release 2. *Astronomy & Astrophysics*, *616*, A1. doi:10.1051/0004-6361/201833051

Chollet, F. (2015). *Keras*. GitHub repository: https://github.com/fchollet/keras

de Bruijne, J. H. J. (2012). Astrometric satellite missions. In EAS Publications Series (Vol. 45, pp. 105–116). doi:10.1051/eas/1245003

He, K., Zhang, X., Ren, S., & Sun, J. (2016). Deep residual learning for image recognition. In *Proceedings of the IEEE conference on computer vision and pattern recognition (CVPR)* (pp. 770-778). IEEE.

Kingma, D. P., & Ba, J. (2014). Adam: A method for stochastic optimization. *arXiv preprint arXiv:1412.6980.*

LeCun, Y., Bengio, Y., & Hinton, G. (2015). Deep learning. *Nature*, *521*(7553), 436–444. doi:10.1038/nature14539 PMID:26017442

Lépine, S., & Shara, M. M. (2005). A Catalog of Northern Stars with Annual Proper Motions Larger than 0.15" (LSPM-NORTH Catalog). *The Astronomical Journal*, *129*(3), 1483–1522. doi:10.1086/427854

Perryman, M. A. C. (2001). The Hipparcos and Tycho catalogues. Astrometric and photometric star catalogues derived from the ESA Hipparcos Space Astrometry Mission. *Astronomy & Astrophysics*, *369*, 339–363. doi:10.1051/0004-6361:20010085

Poduval, B., Pitman, K. M., Verkhoglyadova, O., & Wintoft, P. (2023). Editorial: Applications of statistical methods and machine learning in the space sciences. *Frontiers in Astronomy and Space Sciences*, *10*, 1163530. doi:10.3389/fspas.2023.1163530

Simonyan, K., & Zisserman, A. (2014). Very deep convolutional networks for large-scale image recognition. *arXiv preprint arXiv:1409.1556*.

Srivastava, N., Hinton, G., Krizhevsky, A., Sutskever, I., & Salakhutdinov, R. (2014). Dropout: A simple way to prevent neural networks from overfitting. *Journal of Machine Learning Research*, *15*(1), 1929–1958.

Ting, Y. S., & Ranganath, S. (2019). Machine Learning for Astronomy. *Annual Review of Statistics and Its Application*, *6*, 399–426. doi:10.1146/annurev-statistics-030718-105348

Chapter 9
Synergizing Aerospace Efficiency:
Blockchain and AI Integration for Enhanced Security in Flight Data Management

S. Surya
Saveetha Engineering College, India

R. Elakya
Sri Venkateswara College of Engineering, India

S. Selvanayaki
Saveetha Engineering College, India

ABSTRACT

Blockchain and artificial intelligence are two emerging technologies with a wide range of industrial uses. Artificial intelligence (AI) has the capacity to learn on its own, which makes it useful for automating some industrial tasks. However, blockchain offers decentralized system features to guarantee auditability, security, and transparency of transactions. The aerospace engineering field can benefit from the integration of blockchain technology and artificial intelligence (AI) to improve productivity, security, and transparency across the board. The area of flight data monitoring is also addressed, with a focus on how blockchain protects vital flight data and AI algorithms examine large datasets to improve flying performance, route optimization, and fuel economy. This chapter further investigates the possibilities of using AI algorithms in conjunction with decentralized air traffic management systems made possible by blockchain technology, which can dynamically adjust to changing circumstances in real time to minimize delays and congestion.

DOI: 10.4018/979-8-3693-1491-3.ch009

1. INTRODUCTION

Aerospace engineering, a specialized discipline, is concerned with designing and developing aircraft and systems for commercial and military applications. The demands for unparalleled efficiency and precision in aerospace operations necessitate the deployment of cutting-edge technologies. Despite significant technological advances, an ongoing requirement for enhanced security and reliability persists to ensure operational efficiency and safety in this sector .

Flight data management involves the collection, storage, and analysis of crucial information related to aircraft operations. Ensuring the security and integrity of this data is paramount for the safety of passengers and the overall efficiency of the aviation industry. This section explores how the integration of innovative technologies like Blockchain and AI is transforming security measures in flight data management (Dan et al., 2020). The aerospace industry stands at the forefront of technological innovation, continually pushing the boundaries of efficiency and safety. As aviation systems become increasingly interconnected and data-centric, the need for robust security measures in flight data management has never been more critical (Huynh-The et al., 2023). This chapter explores the transformative synergy of Blockchain and Artificial Intelligence (AI) in elevating security standards within flight data management, ushering in a new era of efficiency, transparency, and resilience and discusses the challenges.

A. Challenges in Flight Data Security

The aviation industry faces numerous challenges in securing flight data:

Cyber Threats: The increasing connectivity of aircraft systems exposes them to potential cyber threats.

Data Tampering: Unauthorized access and tampering of flight data can compromise the accuracy of records.

Real-time Threat Detection: Identifying and responding to security threats in real-time is a complex task.

B. The Role of Blockchain in Flight Data Security

Blockchain technology provides a decentralized and immutable ledger, addressing several security concerns:

Tamper-Proof Records: Blockchain's immutability ensures that once data is recorded, it cannot be altered, providing a secure and transparent historical record of flight data.

Decentralization: By distributing copies of the ledger across a network of nodes, Blockchain reduces the risk of a single point of failure, enhancing resilience against cyber attacks.

Smart Contracts: Automated and self-executing smart contracts embedded in Blockchain can streamline and secure processes, such as maintenance record validation and supply chain management (Abdulrahman et al., 2022; Crosby et al., 2016).

C. The Contribution of Artificial Intelligence

AI plays a crucial role in enhancing security measures and risk management in flight data:

Anomaly Detection: AI algorithms can analyze vast amounts of data in real-time, identifying abnormal patterns or behaviors that may indicate a security threat (Lv et al., 2021; Mattioli et al., 2020).

Predictive Analytics: By leveraging machine learning, AI can predict potential security issues, allowing for proactive measures to be taken before they escalate.

Behavioral Biometrics: AI can incorporate behavioral biometrics for user authentication, ensuring that only authorized personnel access sensitive flight data (Ehsani, 2018).

D. Real-time Monitoring and Threat Identification

Continuous Surveillance: AI-powered systems continuously monitor flight data streams, providing real-time surveillance and analysis.

Threat Identification: The combination of AI's analytical capabilities and Blockchain's secure data storage enables the swift identification and isolation of potential threats (Vaiyapuri et al., 2023).

2. RELATED WORKS

Blockchain technology, renowned for its secure and immutable distributed database, presents a compelling solution for enhancing aerospace security through robust auditing and record-keeping mechanisms. The decentralized nature of blockchain fosters trust and transparency, critical elements in the realm of aerospace security (Al-Breiki et al., 2020; Amari et al., 2023). With its proven ability to maintain data integrity and efficiently record transactions, blockchain technology has found diverse applications within the aerospace sector (Corea, 2019; Reyna et al., 2018).

The transformative potential of blockchain in aerospace security lies in its distinctive architecture and innovative data storage methods. Each data block, securely linked with others through cryptographic hashes, forms a chronologically time-stamped chain (Martintoni et al., 2022; Mondragon et al., 2018). Metadata and header information of these blocks are securely embedded within the blockchain, ensuring a tamper-resistant record. As the aerospace blockchain user base expands, the number of data blocks increases, reinforcing the system's resilience against unauthorized data manipulation.

Blockchain technology employs various transaction validation approaches crucial for aerospace security. The proof-of-work method, exemplified in systems like Bitcoin, involves solving complex mathematical problems to add a new data block to the aerospace blockchain. However, this process, while effective, is energy-intensive and demands substantial computational resources (Crosby et al., 2016). An alternative approach, proof of stake, allocates computational demands proportionately to the data block's size, though widespread implementation is still in progress. Additional algorithms like the Byzantine-fault-tolerance algorithm and the Quorum Slicing algorithm play integral roles in data validation within aerospace blockchain networks. As blockchain technology gains increased traction in aerospace security applications, its evolution aligns with the growing significance of artificial intelligence (AI). AI, a subfield of computer science, has been pivotal for over three decades in drawing meaningful inferences from data. Employing methodologies such as fuzzy logic and support vector machines, AI finds critical applications in aerospace security, ranging from advanced search algorithms to robotics. A milestone in the evolution of AI is the advent of deep learning, which leverages interconnected neural networks to enhance machine learning performance. Foundational AI algorithms and principles, established in the 1980s, continue to shape the landscape of aerospace security. As the aerospace industry increasingly integrates blockchain technology for enhanced security, the continued progress of AI assumes growing importance in fortifying and advancing the sector's overall safety measures.AI has undergone substantial evolution, establishing itself as a powerful technology with the potential to revolutionize the aerospace industry and drive societal changes. Leveraging its capabilities to extract valuable insights from extensive datasets and undertake complex tasks, AI is well-positioned to transform various facets of aerospace, including research, development, and operational processes. The ongoing advancements in computational technologies, combined with the introduction of cutting-edge AI methodologies, point towards an expanding realm of AI applications within aerospace. This growth is poised to catalyze further progress and breakthroughs in the field, ushering in a future marked by continuous innovation and transformative possibilities for the aerospace sector.

3. AI USE CASES AND POTENTIAL USE CASES IN AEROSPACE ENGINEERING

In the realm of aerospace engineering, the integration of artificial intelligence (AI) in flight data management has ushered in transformative capabilities and possibilities. One pivotal application lies in predictive maintenance, where AI algorithms meticulously analyze historical flight data, sensor readings, and maintenance records to prognosticate potential component failures. This not only facilitates proactive maintenance interventions but also substantially minimizes downtime, ultimately bolstering the overall reliability of aircraft. Another critical use case involves anomaly detection and fault isolation, where AI systems, in real-time, scrutinize flight data to swiftly identify irregularities and pinpoint potential faults, thereby enhancing safety measures (Ezzat et al., 2021). Furthermore, AI's prowess is harnessed in fuel efficiency optimization, as it processes intricate data sets encompassing flight data, weather conditions, and air traffic to recommend and implement more fuel-efficient flight paths. This not only contributes to economic savings but also aligns with the aerospace industry's commitment to environmental sustainability. As aerospace engineering advances, the symbiotic relationship between AI and flight data management continues to unfold, promising further innovations that redefine the efficiency, safety, and sustainability standards of the aerospace sector (Henderson et al., 2022; James & Roper, 2018; Martin & Freeland, 2021).

4. CHALLENGES OF BLOCKCHAIN AND AI IN AEROSPACE ENGINEERING FOR ENHANCED SECURITY IN FLIGHT DATA MANAGEMENT

Integrating blockchain and artificial intelligence (AI) into aerospace engineering for enhanced security in flight data management poses its own set of challenges. Here are some key considerations:

Table 1. Challenges of blockchain and AI in aerospace

Key Features	Challenge	Solution
Data Interoperability	Ensuring seamless integration and interoperability between different systems, both within and across organizations, is crucial. Blockchain and AI solutions need to work cohesively to manage flight data securely.	Standards and protocols should be established to facilitate data exchange between various aerospace systems and technologies.
Security and Privacy Concerns	The sensitivity of flight data raises concerns about security and privacy. Blockchain and AI systems must be designed to prevent unauthorized access and protect sensitive information (Ghimire et al., 2022).	Implement robust encryption mechanisms, access controls, and compliance with data protection regulations to ensure the security and privacy of flight data.
Scalability	As the volume of flight data increases, the scalability of blockchain and AI systems becomes crucial. Ensuring that the technology can handle a growing amount of data without compromising performance is a significant challenge.	Design scalable architectures and leverage technologies that support efficient scaling, such as sharding in blockchain and distributed computing in AI.
Integration with Existing Systems	Aerospace systems often have legacy components and infrastructure. Integrating blockchain and AI technologies with existing systems without disrupting operations can be a complex task.	Develop integration strategies that allow for a phased implementation, ensuring compatibility with legacy systems while gradually introducing new technologies.
Regulatory Compliance	The aerospace industry is highly regulated, and compliance with aviation standards and regulations is paramount. Implementing blockchain and AI solutions that adhere to these regulations can be challenging.	Work closely with regulatory bodies to understand and comply with industry standards. Develop solutions that facilitate transparency and traceability, making it easier to demonstrate compliance.
Training and Skillset	The successful implementation of blockchain and AI technologies requires a workforce with the necessary skills and expertise. Training existing personnel and attracting new talent can be a challenge.	Invest in training programs and collaborations with educational institutions to build a skilled workforce. Foster a culture of continuous learning and adaptation to new technologies.
Costs and Return on Investment	Implementing blockchain and AI solutions can involve significant upfront costs. Ensuring a positive return on investment over the long term is essential for justifying these expenses.	Conduct thorough cost-benefit analyses, focusing on both short-term gains and long-term benefits. Identify key performance indicators (KPIs) to measure the success of the implemented technologies.

Some of these challenges include manufacturing digitalisation, weaker supply chains, cyber security, increase in passenger traffic, and rigorous modernisation of existing aircrafts.Addressing these challenges requires a strategic and collaborative approach, involving key stakeholders such as aerospace engineers, data scientists, regulatory bodies, and technology providers (Kavasidis et al., 2023). Continuous innovation, research, and industry-wide collaboration will be essential for successfully enhancing security in flight data management using blockchain and AI in aerospace engineering.

5. AI AND BLOCKCHAIN BENEFITS IN AEROSPACE

Figure 1. AI and blockchain potential benefits in the aerospace industry

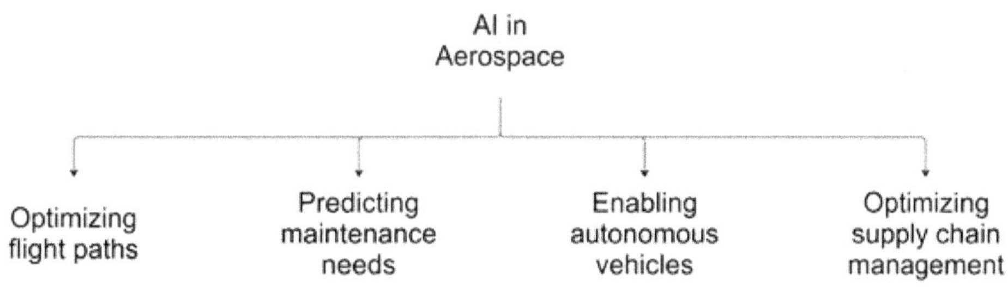

Figure 1 visually encapsulates AI's critical advantages to the aerospace sector, illustrating its multifaceted applications. These range from predictive maintenance and fault detection to optimizing production schedules and implementing smart manufacturing processes. By graphically portraying these benefits, Figure 1 provides an all-encompassing understanding of AI's capacity to augment efficiency, fortify safety measures, and enhance decision-making processes in aerospace engineering and management.

Table 2. Benefits of blockchain and AI in aerospace

Benefits	Blockchain	AI
Data Privacy and Security	While blockchain is known for its security features, ensuring the privacy and security of sensitive flight data stored on the blockchain is critical (Mandolla et al., 2019). Encryption and proper access controls must be implemented to prevent unauthorized access.	AI algorithms often require large datasets for training, and handling sensitive flight data while maintaining privacy is a challenge. Implementing techniques like federated learning (training models on decentralized data) can be explored.
Interoperability	Achieving interoperability between different blockchain platforms and legacy systems used in aerospace is challenging. Standardization efforts are needed to ensure seamless integration and communication between diverse systems.	Integrating AI algorithms with existing avionics systems and databases may require substantial modifications. Compatibility and communication protocols need to be standardized to ensure interoperability.
Regulatory Compliance	Adhering to aviation regulations while utilizing blockchain technology is crucial. Compliance with data protection laws and aviation standards, such as those set by the International Civil Aviation Organization (ICAO), is essential.	The use of AI in aerospace must comply with strict safety and certification standards. Ensuring that AI algorithms meet regulatory requirements, especially in safety-critical systems, is a significant challenge.
Scalability	As the volume of flight data increases, scalability becomes a concern. Blockchain networks must be able to handle a growing number of transactions without compromising performance.	Training AI models on large datasets can be computationally intensive. Developing scalable AI solutions that can handle the increasing amount of data generated in aerospace operations is essential.
Data Accuracy and Integrity	While blockchain ensures data integrity once a record is added, ensuring the accuracy of the initial data input is crucial. Garbage in, garbage out (GIGO) remains a concern.	AI models are susceptible to biases present in training data. Ensuring the accuracy and fairness of AI predictions requires careful curation of training datasets and ongoing monitoring.

6. LIMITATIONS

Aerospace systems demand high reliability and fault tolerance. Ensuring that blockchain and AI systems can operate seamlessly under adverse conditions, such as hardware failures or network disruptions, is crucial for maintaining the integrity of flight data. Many aerospace systems rely on legacy infrastructure. Integrating blockchain and AI technologies into existing systems without disrupting operations or compromising safety is a significant challenge. Compatibility issues with older technologies may arise during the integration process. Flight data generates large volumes of information, and real-time processing is crucial for safety and decision-making. Blockchain networks may face limitations in terms of transaction throughput and data storage, potentially causing delays in data processing (Gunasekara et al., 2022). As the number of flights and aircraft increases, the blockchain network must scale to accommodate the growing amount of data. Scalability issues can lead to slower transaction speeds and increased latency, impacting the real-time nature of flight data management.

Blockchain transactions involve consensus mechanisms that may introduce latency. In aviation, where split-second decisions are critical, delays in data validation and processing could pose a significant challenge. Smart contracts, which execute automatically when predefined conditions are met, are integral to many blockchain applications. However, vulnerabilities in smart contracts could lead to security breaches or unintended consequences, posing a risk to flight data integrity. While blockchain provides transparency, there may be concerns regarding the privacy and confidentiality of sensitive flight data. Striking the right balance between transparency and data security is crucial to address privacy issues (Chang & Svetinovic, 2020; Gunasekara et al., 2022).

7. CONCLUSION

In conclusion, the integration of blockchain technology and artificial intelligence presents a promising avenue for advancing the aerospace engineering field. The combination of AI's autonomous learning capabilities and blockchain's decentralized features holds the potential to revolutionize various aspects of the industry, ranging from enhancing productivity and ensuring security to providing unprecedented transparency in transactions. The focus on flight data monitoring underscores the critical role of these technologies in safeguarding vital information. Blockchain's ability to guarantee auditability and security complements AI algorithms' capacity to analyze large datasets, contributing to improved flying performance, route optimization, and enhanced fuel economy. This integration not only addresses current challenges but also sets the stage for future innovations in aerospace engineering. Moreover, the exploration of decentralized air traffic management systems showcases the dynamic adaptability facilitated by blockchain technology. By harnessing AI algorithms in conjunction with a decentralized framework, the industry can proactively respond to changing circumstances in real time, minimizing delays and congestion. This not only enhances operational efficiency but also contributes to a safer and more sustainable air travel ecosystem. In essence, the synergistic integration of blockchain and AI opens new frontiers for innovation in aerospace engineering. As the industry continues to evolve, embracing these technologies holds the promise of ushering in an era of heightened productivity, unprecedented security, and unparalleled transparency. The potential benefits outlined in this chapter underscore the transformative impact that the convergence of blockchain and AI can have on shaping the future of air travel and aerospace operations.

REFERENCES

Abdulrahman, Y., Parezanovic, V., & Svetinovic, D. (2022). AI-blockchain systems in aerospace engineering and management: Review and challenges. *Proc. 30th Telecommun. Forum (TELFOR),* 1–4. 10.1109/TELFOR56187.2022.9983700

Al-Breiki, H., Rehman, M. H. U., Salah, K., & Svetinovic, D. (2020). Trustworthy blockchain oracles: Review, comparison, and open research challenges. *IEEE Access : Practical Innovations, Open Solutions*, *8*, 85675–85685. doi:10.1109/ACCESS.2020.2992698

Amari, H., Houda, Z. A. E., Khoukhi, L., & Belguith, L. H. (2023). Trust management in vehicular ad-hoc networks: Extensive survey. *IEEE Access : Practical Innovations, Open Solutions*, *11*, 47659–47680. doi:10.1109/ACCESS.2023.3268991

Chang, T.-H., & Svetinovic, D. (2020, January). Improving Bitcoin ownership identification using transaction patterns analysis. *IEEE Transactions on Systems, Man, and Cybernetics. Systems*, *50*(1), 9–20. doi:10.1109/TSMC.2018.2867497

Corea, F. (2019). The convergence of AI and blockchain. In *Applied Artificial Intelligence: Where AI Can Be Used.* Springer., doi:10.1007/978-3-319-77252-3_4

Crosby, M., Pattanayak, P., Verma, S., & Kalyanaraman, V. (2016). Blockchain technology: Beyond bitcoin. *Appl. Innov.*, *2*(6–10), 71.

Dan, W., Yin, L., Yeying, W., Rui, W., Chen, F., Zhongyang, Z., Dongdong, H., & Liang, G. (2020). Research on aerospace data transaction platform and method based on blockchain. *Proc. 2nd Int. Conf. Inf. Technol. Comput. Appl. (ITCA),* 410–413. 10.1109/ITCA52113.2020.00092

Ehsani, F. (2018). *Blockchain in finance: From buzzword to watchword in 2016.* Available: https://www.coindesk.com/markets/2016/12/20/blockchain-in-fi nancefrom-buzzword-to-watchword-in-2016/

Ezzat, D., Hassanien, A. E., Darwish, A., Yahia, M., Ahmed, A., & Abdelghafar, S. (2021). Multi-objective hybrid artificial intelligence approach for fault diagnosis of aerospace systems. *IEEE Access : Practical Innovations, Open Solutions*, *9*, 41717–41730. doi:10.1109/ACCESS.2021.3064976

Ghimire, T., Joshi, A., Sen, S., Kapruan, C., Chadha, U., & Selvaraj, S. K. (2022, January). Blockchain in additive manufacturing processes: Recent trends & its future possibilities. *Materials Today: Proceedings*, *50*, 2170–2180. doi:10.1016/j.matpr.2021.09.444

Gunasekara, H. G., Sridarran, P., & Rajaratnam, D. (2022, May). Effective use of blockchain technology for facilities management procurement process. *Journal of Facilities Management, 20*(3), 452–468. doi:10.1108/JFM-10-2020-0077

Henderson, A., Harbour, S., & Cohen, K. (2022). Toward airworthiness certification for artificial intelligence (AI) in aerospace systems. *Proc. IEEE/AIAA 41st Digit. Avionics Syst. Conf. (DASC),* 1–10. 10.1109/DASC55683.2022.9925740

Huynh-The, T., Gadekallu, T. R., Wang, W., Yenduri, G., Ranaweera, P., Pham, Q.-V., da Costa, D. B., & Liyanage, M. (2023, June). Blockchain for the metaverse: A review. *Future Generation Computer Systems, 143*, 401–419. doi:10.1016/j.future.2023.02.008

James, T., & Roper, S. (2018). *Launching from earth: The science behind space law and technological developments. In Deep Space Commodities.* Palgrave Macmillan. doi:10.1007/978-3-319-90303-3_3

Kavasidis, I., Lallas, E., Mountzouris, G., Gerogiannis, V. C., & Karageorgos, A. (2023). A federated learning framework for enforcing traceability in manufacturing processes. *IEEE Access : Practical Innovations, Open Solutions, 11*, 57585–57597. doi:10.1109/ACCESS.2023.3282316

Lv, Z., Chen, D., Lou, R., & Alazab, A. (2021, April). Artificial intelligence for securing industrial-based cyber–physical systems. *Future Generation Computer Systems, 117*, 291–298. doi:10.1016/j.future.2020.12.001

Mandolla, C., Petruzzelli, A. M., Percoco, G., & Urbinati, A. (2019, August). Building a digital twin for additive manufacturing through the exploitation of blockchain: A case analysis of the aircraft industry. *Computers in Industry, 109*, 134–152. doi:10.1016/j.compind.2019.04.011

Martin, A.-S., & Freeland, S. (2021). The advent of artificial intelligence in space activities: New legal challenges. *Space Policy, 55*(Feb), 101408. doi:10.1016/j.spacepol.2020.101408

Martintoni, D., Senni, V., Marin, E. G., & Gutierrez, A. J. C. (2022). Sensitive information protection in blockchain-based supply-chain management for aerospace. *Proc. IEEE Int. Conf. Omni-layer Intell. Syst. (COINS),* 1–8. 10.1109/COINS54846.2022.9854974

Mattioli, J., Perico, P., & Robic, P.-O. (2020). Improve total production maintenance with artificial intelligence. *Proc. 3rd Int. Conf. Artif. Intell. Ind. (AI4I),* 56–59. 10.1109/AI4I49448.2020.00019

Mondragon, A. E. C., Mondragon, C. E. C., & Coronado, E. S. (2018). Exploring the applicability of blockchain technology to enhance manufacturing supply chains in the composite materials industry. *Proc. IEEE Int. Conf. Appl. Syst. Invention (ICASI),* 1300–1303. 10.1109/ICASI.2018.8394531

Reyna, A., Martín, C., Chen, J., Soler, E., & Díaz, M. (2018, November). On blockchain and its integration with IoT. Challenges and opportunities. *Future Generation Computer Systems*, *88*, 173–190. doi:10.1016/j.future.2018.05.046

Vaiyapuri, T., Shankar, K., Rajendran, S., Kumar, S., Acharya, S., & Kim, H. (2023). Blockchain assisted data edge verification with consensus algorithm for machine learning assisted IoT. *IEEE Access : Practical Innovations, Open Solutions*, *11*, 55370–55379. doi:10.1109/ACCESS.2023.3280798

Chapter 10
Synergizing AI and Blockchain:
Transforming Aerospace Engineering Operations

R. Elakya
Sri Venkateswara College of Engineering, India

T. Manoranjitham
SRM Institute of Science and Technology, India

R. Thanga Selvi
Vel Tech Rangarajan Dr. Sagunthala R&D Institute of Science and Technology, India

S. Shanthana
Rajalakshmi Institute of Technology, Chennai, India

ABSTRACT

The aerospace industry stands on the brink of transformative advancements with the convergence of artificial intelligence (AI) and blockchain technologies. The chapter commences by highlighting the imperative role of AI and blockchain in the aerospace sector. Subsequently, the chapter delves into the distinct applications of AI within aerospace engineering. Autonomous systems, empowered by AI, are explored for their influence on unmanned aerial vehicles (UAVs), spacecraft navigation, and decision-making. The integration of AI-driven predictive maintenance algorithms emerges as a pivotal strategy to optimize maintenance schedules, reduce downtime, and augment safety through accurate prognostics. Further, the utilization of AI for flight path optimization is detailed, showcasing its ability to reduce fuel consumption and minimize environmental impact. The role of AI extends to the realms of aerodynamics, design, and materials science, accelerating the creation of innovative aircraft and spacecraft designs.

DOI: 10.4018/979-8-3693-1491-3.ch010

INTRODUCTION

The aerospace industry stands as a paragon of innovation, where precision, efficiency, and safety are paramount. As the industry constantly seeks ways to push the boundaries of technological advancement, the convergence of two groundbreaking technologies—Artificial Intelligence (AI) and blockchain—holds the promise of a paradigm shift. This introduction embarks on a journey to explore the profound implications of integrating AI and blockchain within the aerospace engineering domain.

In recent years, AI has emerged as a formidable force, underpinned by its capacity to analyze vast datasets, learn autonomously, and make informed decisions. In the aerospace sector (Lin & Liao, 2017), AI offers an array of capabilities, from enhancing predictive maintenance models to optimizing flight paths, driving innovation in aircraft design, and even revolutionizing materials science. Its potential to usher in unprecedented operational efficiency and elevate safety standards is undeniable.

Simultaneously, blockchain, characterized by its transparent, immutable, and secure ledger, has come to the forefront. Within the aerospace industry (Lohmer & Lasch, 2020), its decentralized structure can dramatically improve supply chain management, secure the integrity of maintenance records, and facilitate the sharing of critical data across a complex network of stakeholders. It underpins trust, transparency, and accountability, making it a transformative technology.

However, the true transformative power emerges when these two technologies converge. AI (Longo et al., 2019), with its analytical prowess, can generate insights from vast datasets. Blockchain, with its secure and tamper-proof record-keeping, provides the infrastructure to record these insights transparently. The fusion of AI-generated insights and blockchain's data integrity capabilities creates a potent synergy (Madhwal & Panfilov, 2017), enhancing decision-making processes, ensuring regulatory compliance, and driving aerospace engineering into a new era.

This chapter embarks on a comprehensive exploration of this synergistic integration of AI and blockchain in aerospace engineering. It will dissect the individual applications of AI and blockchain (Makridakis et al., 2018), uncovering how they independently address critical challenges in the industry. Subsequently, the chapter will delve into the fusion of these technologies, illustrating how they complement each other to create a holistic solution. Real-world case studies and practical frameworks will exemplify their transformative potential, while challenges and future trends will be scrutinized.

In the aerospace industry's unending quest for efficiency, safety, and innovation, AI and blockchain offer an opportunity to rewrite the rules (Malavolta et al., 2019). This introduction sets the stage for an in-depth examination of the convergence of

AI and blockchain, propelling aerospace engineering into a future of unprecedented possibilities.

AI (Artificial Intelligence) and blockchain are two innovative technologies that can be leveraged to optimize various aspects of aerospace engineering (Mammadzada et al., 2020). Let's explore how each technology can be applied in this context:

1. AI IN AEROSPACE ENGINEERING

AI involves the development of systems that can perform tasks that typically require human intelligence (Rana et al., 2019). In aerospace engineering, AI can be utilized in several ways:

- **Autonomous Systems:** AI-powered autonomous systems are a transformative force in aerospace engineering, enhancing navigation, decision-making, and collision avoidance for a variety of aerial and space vehicles (Liu et al., 2020). Drones, satellites, and aircraft benefit from AI-driven navigation that adapts to changing conditions, making dynamic decisions based on sensor data. These systems are critical for safety, particularly in scenarios with limited human intervention, such as remote planetary exploration or unmanned surveillance missions. In UAVs, real-time collision avoidance capabilities enable drones to adjust their flight path to avoid obstacles, ensuring safety in diverse applications (Niculescu-Mizil, 2020). Space exploration relies on AI to navigate complex environments, avoiding collisions with debris and celestial bodies. AI's autonomy is vital for spacecraft, allowing them to make decisions independently when communication with Earth is subject to significant delays. This convergence of AI and autonomous systems propels aerospace engineering into a new era of efficiency and safety, redefining the boundaries of innovation in this field.
- **Predictive Maintenance:** Predictive maintenance (Chuck et al., 2019), driven by AI algorithms, is a pivotal advancement in aerospace engineering, revolutionizing the management of aircraft and spacecraft. By meticulously scrutinizing sensor data from critical components and engines, AI can accurately predict maintenance needs. This proactive approach minimizes unexpected downtime, leading to greater operational efficiency and heightened safety as potential issues are addressed before they become critical. In addition to its safety and efficiency gains, predictive maintenance delivers substantial cost savings by optimizing maintenance schedules, reducing labor and material expenses, and extending the longevity and reliability of aerospace components. In essence, AI-enabled predictive maintenance is a

cornerstone of a more dependable, cost-effective, and sustainable aerospace industry.

- **Flight Path Optimization:** Flight path optimization, powered by AI algorithms, marks a crucial leap in aerospace efficiency and sustainability. By processing vast datasets encompassing weather conditions, air traffic, and various parameters, AI can meticulously fine-tune flight paths for aircraft, resulting in reduced fuel consumption and emissions. This optimization not only benefits airlines' bottom lines but also plays a pivotal role in mitigating the environmental impact of aviation. By minimizing fuel consumption, AI-driven flight path optimization aligns with the imperative for sustainable aviation practices, striving to make air travel more economically and ecologically viable while meeting the ever-increasing demand for air transportation.
- **Aerodynamics and Design:** AI's application in aerospace engineering extends to aerodynamics and design, where it revolutionizes the way aircraft and spacecraft are conceptualized and developed. By employing AI-driven simulations of complex aerodynamic models, engineers can rapidly explore a multitude of design iterations, saving significant time and resources. This accelerates the design process, enabling the creation of vehicles that are not only innovative but also highly efficient. AI's ability to process extensive datasets and perform complex calculations swiftly empowers engineers to make informed decisions, resulting in aircraft and spacecraft designs that are more aerodynamically sound, safer, and environmentally friendly. This approach fosters the creation of next-generation aerospace vehicles that push the boundaries of efficiency, performance, and sustainability, marking a profound shift in aerospace design practices.
- **Materials Science:** AI's role in materials science within the aerospace industry is transformative, offering an expedited pathway to the discovery and design of advanced materials. By leveraging machine learning and data analysis, AI can sift through vast datasets to identify material properties, compositions, and structures that would be challenging or impossible to uncover through traditional methods. This accelerates the development of novel materials that are not only lighter but also stronger and more efficient, addressing critical aerospace requirements. These materials play a pivotal role in enhancing aircraft and spacecraft performance, reducing weight, improving fuel efficiency, and ensuring structural integrity. AI's contribution to materials science reshapes the aerospace landscape, propelling the industry into an era of innovation and efficiency, where advanced materials are at the forefront of design and development.

2. BLOCKCHAIN IN AEROSPACE ENGINEERING

Blockchain is a distributed and tamper-proof digital ledger technology that offers transparency, security, and traceability. In aerospace engineering, blockchain can have various applications:

- **Supply Chain Management:** Supply chain management in the aerospace industry is inherently intricate, often involving a vast network of suppliers and partners responsible for delivering crucial components to build and maintain aircraft and spacecraft. Blockchain technology offers an innovative and robust solution to address the challenges within this multifaceted supply chain.

Blockchain serves as a transparent and secure ledger, meticulously recording every transaction and movement of aerospace components from their origin to their final destination. This end-to-end visibility is invaluable, as it ensures that the authenticity and provenance of every part can be thoroughly verified at each stage. This transparency reduces the risk of counterfeit or substandard components infiltrating the supply chain, a critical concern in an industry where safety and reliability are paramount.

Moreover, the secure and tamper-proof nature of blockchain ensures the integrity of records. Once data is recorded, it becomes immutable, making it virtually impossible to manipulate or falsify information. This level of trust and security adds a layer of accountability to the aerospace supply chain, instilling confidence among stakeholders and regulatory authorities. The application of blockchain in aerospace supply chain management redefines how the industry ensures the authenticity and integrity of components. It offers a transparent, secure, and traceable solution to a complex and critical challenge, reducing the risk of counterfeit parts and enhancing the overall efficiency, trustworthiness, and accountability within the aerospace supply chain.

- **Maintenance and Repair:** Maintenance and repair operations are critical in ensuring the safety and airworthiness of aircraft and spacecraft in the aerospace industry. Blockchain technology plays a transformative role in this domain by providing a secure and transparent ledger for recording and storing maintenance records, repair history, and component performance data. Here's a detailed explanation of how blockchain improves maintenance and repair processes in aerospace engineering:
  1. **Transparent Maintenance Records:** Blockchain offers a decentralized ledger that records every maintenance activity, from routine check-ups to extensive repairs. Each entry is time-stamped and linked to the specific

component or system it pertains to. This transparency ensures that the maintenance history is clear, accessible, and tamper-proof, reducing the risk of falsified or incomplete records.

2. **Comprehensive Repair History:** In addition to maintenance records, blockchain captures detailed repair histories. It documents the nature of each repair, the parts used for replacement, and the technicians responsible for the work. This comprehensive repair history is invaluable for analyzing the performance of maintenance teams, tracking recurring issues, and planning for future maintenance needs.
3. **Component Performance Data:** Blockchain also securely records component performance data over time. This includes information about the health and reliability of critical components. By maintaining a detailed performance history, aerospace engineers can make informed decisions about component replacement or refurbishment. Moreover, the data can be leveraged with artificial intelligence (AI) and analytics to predict maintenance needs, preventing unexpected component failures and optimizing maintenance schedules.
4. **Verification and Compliance:** The ability to verify the maintenance history of critical components is a significant advantage of blockchain in aerospace engineering. Regulatory authorities and aviation safety agencies can readily access and audit the maintenance and repair records. This transparency contributes to safety and regulatory compliance, ensuring that maintenance activities align with stringent aerospace standards and requirements. Blockchain's application in storing maintenance and repair data enhances safety, efficiency, and compliance in the aerospace industry. It provides a secure, transparent, and tamper-proof solution for verifying the maintenance history of critical components, ensuring that aerospace vehicles remain in optimal condition and meet rigorous safety and compliance standards.

- **Data Sharing:** In the collaborative landscape of aerospace engineering, data sharing is a critical aspect that often involves numerous organizations, including manufacturers, operators, regulatory bodies, and maintenance providers. Blockchain technology offers a transformative solution to streamline and secure data sharing while maintaining data integrity and access control. Here's a comprehensive explanation of how blockchain facilitates data sharing in aerospace engineering:

1. **Secure Data Sharing:** Blockchain creates a secure, decentralized ledger where data can be shared among authorized participants. This security is achieved through encryption and cryptographic techniques, making it exceptionally difficult for unauthorized access or tampering to occur. This

level of security is paramount in aerospace, where sensitive information, such as design specifications, maintenance records, and safety data, must be protected from cyber threats.

2. **Data Integrity:** Once data is recorded on the blockchain, it becomes immutable and transparent. This immutability ensures that data remains unchanged and tamper-proof, providing a high level of trust in the accuracy and authenticity of the shared information. This feature is vital for critical aerospace data, as it guarantees that data integrity is maintained throughout the data's lifecycle.
3. **Access Control:** Blockchain's access control mechanisms enable organizations to define who can access specific information on the blockchain. This granular control over data access is advantageous when different stakeholders need access to different parts of the data. For example, aircraft manufacturers might grant access to maintenance records to authorized maintenance providers while withholding sensitive design data.
4. **Smart Contracts:** Smart contracts, which are self-executing code embedded within the blockchain, can automate data-sharing processes. These contracts can be set up to trigger actions when predefined conditions are met. In aerospace, this can mean automatically sharing maintenance records with regulatory authorities upon the completion of specific checks or maintenance activities.
5. **Trust and Transparency:** Blockchain's transparency and decentralized nature ensure that all participants can trust the data being shared. Every party involved can verify the authenticity of data on the blockchain independently, reducing the need for third-party verification and enhancing trust among collaborators. Blockchain technology is a game-changer in aerospace engineering by providing a secure, tamper-proof, and transparent platform for data sharing. It enhances data integrity, allows for precise access control, and automates data-sharing processes through smart contracts, fostering efficient collaboration and maintaining the highest standards of data security in an industry where information accuracy and safety are paramount.

- **Flight Data and Safety:** The utilization of blockchain technology for storing flight data and ensuring safety in aerospace engineering is a significant advancement in the industry. Here's a detailed explanation of how blockchain enhances flight data management and safety:
  1. **Immutable Flight Data:** Blockchain creates a secure and immutable ledger to record a wealth of flight data, encompassing sensor readings, performance metrics, and other critical information. Once data is entered

on the blockchain, it becomes unchangeable, ensuring the integrity and accuracy of the flight data over time. This is particularly crucial for accident investigations and maintaining an authentic historical record of each flight.

2. **Accident Investigation:** In the unfortunate event of an accident or incident, the blockchain serves as an indisputable source of flight data. Investigators can access this secure and tamper-proof record to reconstruct the sequence of events and determine the causes of the accident with a high degree of certainty. This aids in preventing future accidents by identifying root causes and improving safety protocols.
3. **Data Accuracy:** Maintaining data accuracy is paramount in aerospace engineering. The immutability of blockchain ensures that flight data remains unaltered and that the information recorded is reliable and accurate. This not only contributes to safety but also enhances decision-making, maintenance planning, and performance analysis.
4. **Improved Safety Protocols:** Blockchain's role in flight data management and safety extends to the enhancement of safety protocols. By providing a secure and transparent platform for recording data, it encourages proactive safety measures, such as real-time monitoring and preventive maintenance. The data stored on the blockchain can trigger automated safety procedures and alerts when predefined thresholds or anomalies are detected.
5. **Regulatory Compliance:** The transparency and immutability of blockchain support compliance with stringent aviation regulations. Regulatory authorities can easily access and audit flight data, ensuring that safety and operational standards are met. This promotes a culture of safety and accountability within the aerospace industry. Blockchain's integration into flight data management in aerospace engineering safeguards the accuracy and integrity of critical data, enhances accident investigations, and fosters proactive safety measures. It contributes to overall safety and reliability within the industry by providing a secure and transparent platform for recording and verifying flight data, serving as an invaluable resource for accident prevention and safety improvement.

- **Smart Contracts:** Blockchain's smart contract functionality is a game-changer in aerospace engineering, automating numerous processes and reducing administrative overhead. Here's an in-depth explanation of how smart contracts enhance efficiency and automation in the aerospace industry:

1. **Aircraft Leasing Agreements:** Smart contracts can automate aircraft leasing agreements. These contracts are self-executing and trigger predefined actions when specific conditions are met. For example, when

lessees make their lease payments, the smart contract can automatically release ownership rights to the aircraft. This reduces the need for intermediaries, streamlining the leasing process and ensuring accuracy.

2. **Insurance Claims:** In the case of insurance claims related to aerospace incidents, smart contracts can facilitate swift and automated claim processing. These contracts can assess the evidence and automatically release funds to the insured parties when predefined conditions for a valid claim are met. This expedites claim settlements and minimizes the potential for disputes, improving the overall efficiency of insurance processes.
3. **Compliance Checks:** Smart contracts are invaluable for automating compliance checks within the aerospace industry. Regulatory compliance is paramount, and smart contracts can ensure that all requirements are met without manual intervention. For instance, they can automate routine checks and documentation verification, ensuring that aircraft and operators adhere to safety and operational standards.
4. **Maintenance Scheduling:** Aerospace maintenance scheduling can benefit from smart contracts as well. These contracts can monitor sensor data and trigger maintenance activities automatically when predefined thresholds are reached. This proactive approach minimizes downtime and ensures that maintenance is conducted precisely when needed, optimizing aircraft performance and safety.
5. **Cost Reduction:** By automating these processes, smart contracts significantly reduce administrative overhead. They eliminate the need for intermediaries and reduce the risk of human error, resulting in cost savings and operational efficiency. This is particularly relevant in an industry where precision and financial prudence are paramount. Smart contracts, empowered by blockchain, revolutionize various processes in aerospace engineering, ranging from leasing agreements to insurance claims and compliance checks. Their automation capabilities reduce administrative burdens, minimize human error, and improve efficiency, ultimately contributing to a more streamlined and cost-effective aerospace industry.

Combining AI and Blockchain in Aerospace:

AI and blockchain can complement each other in aerospace engineering. For instance, AI-powered sensors can collect data from aircraft components, and this data can be securely stored on a blockchain for transparency and traceability. AI algorithms can then analyze this data to predict maintenance needs and optimize performance.

The integration of AI and blockchain in aerospace engineering presents a powerful synergy where these technologies complement each other in innovative ways. Here's a comprehensive explanation of how AI and blockchain combine to enhance the industry:

Data Collection and Security:

In aerospace engineering, a sensor network is a fundamental element of data collection and monitoring, essential for ensuring the safety and performance of aircraft, including both commercial airliners and spacecraft. This network comprises a multitude of advanced sensors that are strategically positioned on critical components and systems throughout the aircraft. Here's a detailed exploration of how these sensor networks function:

Sensor Placement: Aircraft are equipped with a precise and comprehensive sensor network that encompasses crucial components such as engines, avionics, landing gear, airframes, and various onboard systems. The placement of sensors is carefully considered to cover areas that are integral to the aircraft's operation, as well as those that are particularly prone to wear and tear.

Multifaceted Sensors: The sensors in this network are versatile and multifaceted. They are designed to capture a wide array of parameters, ranging from environmental conditions to the performance of mechanical and electronic components. Some common parameters measured by these sensors include:

- Temperature: Sensors monitor temperature variations in critical systems, helping to identify overheating or cooling issues that may require immediate attention.
- Pressure: Pressure sensors gauge changes in atmospheric pressure, cabin pressure, or hydraulic systems to ensure that aircraft pressure is maintained within safe limits.
- Vibration: Vibration sensors detect oscillations or vibrations in components, which can indicate misalignments, wear, or potential mechanical failures.
- Humidity: Sensors measure humidity levels to prevent moisture-related damage and to ensure the comfort of passengers in the cabin.
- Electrical Currents: Electrical sensors monitor the flow of electrical currents within avionics and electrical systems, detecting irregularities or anomalies.

AI-powered sensors are deployed to collect a wealth of data from aircraft components in real-time. This data, which includes information about the condition, performance, and health of critical components, is sensitive and valuable. Blockchain provides a secure and immutable ledger where this data can be stored. This ensures the integrity and confidentiality of the data, safeguarding it against unauthorized access or tampering.

Real-time monitoring in aerospace engineering is a fundamental aspect of sensor networks, where sensors continuously collect data while an aircraft is in operation, providing a constant and up-to-the-minute stream of information. This real-time monitoring is indispensable for swiftly identifying any deviations from normal operating conditions or potential issues. Here's a more detailed explanation of real-time monitoring's significance:

Immediate Data Acquisition: Real-time monitoring entails the immediate collection of data as events unfold. Sensors, strategically placed throughout the aircraft, capture information from critical components and systems without delay. This data is then made instantly available for analysis.

Prompt Issue Detection: The primary objective of real-time monitoring is the early detection of anomalies or deviations from the expected parameters of operation. Any fluctuations, irregularities, or potential issues are identified in their initial stages. This is especially critical for safety, as it allows for the timely initiation of corrective actions before a situation escalates.

Safety Assurance: Real-time monitoring contributes significantly to safety assurance in aerospace. It enables the rapid identification of safety-critical conditions, such as engine overheating, loss of cabin pressure, or control system malfunctions. Detecting these conditions early on helps prevent accidents or incidents, ensuring the well-being of passengers and crew.

Maintenance Efficiency: In addition to safety, real-time monitoring has a direct impact on maintenance efficiency. By continuously assessing the condition of critical components, it aids in identifying maintenance needs precisely when they arise. This proactive approach to maintenance enhances aircraft reliability, reduces downtime, and minimizes operational disruptions.

Data-Driven Decision-Making: Real-time data supports data-driven decision-making. Flight crews, maintenance teams, and operators can rely on the real-time information provided by sensors to make informed decisions. For example, in-flight adjustments or diversions can be planned based on real-time data, contributing to flight safety and efficiency.

Preventive Measures: Beyond immediate corrective actions, real-time monitoring enables the implementation of preventive measures. It helps avoid potential issues by identifying conditions that could lead to problems in the future. For example, real-time data can trigger alerts for maintenance activities to address early signs of component wear or deterioration.

2. TRANSPARENCY AND TRACEABILITY

The data stored on the blockchain is transparent and traceable. Every entry is time-stamped and linked to specific components or systems, creating a comprehensive historical record. This transparency enhances trust among stakeholders, allowing them to independently verify the authenticity and integrity of the data. For regulatory authorities and maintenance teams, this feature is invaluable, ensuring that data remains reliable and tamper-proof. Transparency and traceability are central attributes of blockchain technology when applied to aerospace engineering, significantly enhancing the integrity and accessibility of stored data. Here's a comprehensive explanation of these crucial features:

1. Transparent Data Records: In a blockchain, every data entry, or "block," is openly accessible to authorized participants within the network. This means that all stakeholders involved in aerospace operations, from airlines to regulatory authorities, have visibility into the data recorded on the blockchain. This transparency ensures that the recorded information is not hidden or concealed.

2. Time-Stamped Entries: Each entry in the blockchain is time-stamped, indicating precisely when the data was recorded. This chronological record is essential for tracking the history of components, maintenance activities, and any changes or updates. Time-stamps create a chronological order that is critical for understanding the sequence of events.

3. Component and System Specificity: Blockchain data is meticulously linked to specific components, systems, or aircraft. This means that data is not aggregated into general records but is granular and component-specific. It's possible to trace the performance, maintenance, and history of each component with precision.

4. Comprehensive Historical Record: Over time, the blockchain accumulates a comprehensive historical record of all data entries related to aerospace operations. This historical data offers a valuable resource for tracking the lifecycle of aircraft components, systems, and the aircraft itself. It ensures that past events, maintenance, and operational data are always accessible and verifiable.

5. Independent Data Verification: Transparency allows stakeholders to independently verify the authenticity and integrity of data stored on the blockchain. This independence is essential for building trust among various parties involved in aerospace operations, as it eliminates the need for centralized authorities or intermediaries to validate data.

6. Invaluable for Regulatory Compliance: Regulatory authorities and maintenance teams find the transparency and traceability features of blockchain invaluable. Regulatory bodies can readily access and audit data to ensure compliance with safety, maintenance, and operational standards. Maintenance teams can efficiently verify the maintenance history of components and systems, streamlining maintenance and inspection processes.

7. Tamper-Proof Data: The nature of blockchain technology makes data tampering virtually impossible. Once data is recorded on the blockchain, it is extremely challenging to alter or delete it. This tamper-proof quality ensures that the data remains reliable and authentic throughout its lifecycle.

3. PREDICTIVE MAINTENANCE

AI algorithms are employed to analyze the data stored on the blockchain. These algorithms can process vast datasets and detect patterns and anomalies. By doing so, AI can predict maintenance needs with remarkable accuracy. For instance, AI can identify wear and tear in specific components and forecast when maintenance or replacements are required. This proactive approach reduces downtime, enhances safety, and optimizes maintenance costs.

One commonly used algorithm for predictive maintenance in aerospace engineering is the Random Forest algorithm. The Random Forest algorithm is a popular machine learning algorithm that is particularly well-suited for predictive maintenance and other data-driven applications in aerospace engineering. Let's take a detailed look at how the Random Forest algorithm works:

Step 1: Data Collection and Preprocessing

- Data is collected from various sensors on the aircraft's components and systems, including engine temperature, pressure, and vibration.
- The data is preprocessed, which includes cleaning, normalization, and feature selection to prepare it for analysis.

Step 2: Ensemble Learning

- The Random Forest algorithm is an ensemble learning method, which means it combines multiple decision trees to make more accurate predictions.
- It creates a forest of decision trees, where each tree is trained on a subset of the data and a random subset of features. This randomness ensures diversity and robustness in the model.

Step 3: Training the Model

- Historical data is used to train the Random Forest model. This data contains information about sensor readings, maintenance records, and component failures.
- The algorithm learns to recognize patterns and relationships between sensor data and maintenance needs by building a multitude of decision trees.

Step 4: Pattern Recognition and Anomaly Detection

- Once the model is trained, it can recognize patterns and anomalies in the data. For example, it might detect that a combination of high engine temperature and abnormal vibration levels often precedes a component failure.
- The Random Forest algorithm excels at detecting complex relationships and interactions within the data, making it well-suited for predictive maintenance.

Step 5: Predictive Models

- The algorithm uses the patterns and anomalies it has recognized to make predictions. It can estimate the likelihood of a component failure or maintenance need based on the current sensor data.
- For instance, if it observes a combination of sensor readings that historically led to maintenance, it will predict that maintenance is likely to be needed soon.

Step 6: Real-Time Monitoring

- The algorithm operates in real-time, continuously analyzing new sensor data as it's generated during aircraft operation.
- When it detects patterns or anomalies that are indicative of impending maintenance needs, it generates alerts or notifications.

Step 7: Cost Optimization

- Predictive maintenance based on the Random Forest algorithm allows for cost optimization. Maintenance can be scheduled proactively based on the algorithm's predictions, reducing unscheduled downtime and optimizing maintenance costs.
- The algorithm's ability to recognize complex patterns and relationships within the data enhances the accuracy of predictions, ultimately saving time and resources.

4. PERFORMANCE OPTIMIZATION

Performance optimization is a critical aspect of aerospace engineering, and AI is at the forefront of revolutionizing this field. By harnessing the power of AI algorithms, aerospace engineers can continuously analyze data from a myriad of sensors and sources on aircraft and spacecraft, enabling them to fine-tune various aspects of flight operations. One of the primary contributions of AI in this domain is the ability to optimize flight paths by factoring in dynamic variables such as weather conditions, air traffic, and the specific performance capabilities of the aircraft. This optimization results in more fuel-efficient routes, a reduction in greenhouse gas emissions, and substantial cost savings for airlines, aligning the industry with environmental sustainability goals. Additionally, AI-driven algorithms excel in detecting inefficiencies within aircraft systems and components, enabling the identification of irregularities in engine performance or recommendations for adjustments to improve aerodynamics and reduce drag. Notably, AI doesn't stop at detection but actively suggests improvements, which can be relayed to flight crews or ground-based maintenance teams in real time. Continuous monitoring during flights ensures that adjustments can be made dynamically, ultimately leading to smoother and more efficient flights, enhanced passenger comfort, and reduced wear and tear on aircraft components.

5. DATA DRIVEN DECISION MAKING

The integration of AI and blockchain technologies in the realm of aerospace engineering represents a watershed moment, revolutionizing decision-making through data-driven approaches. This powerful amalgamation empowers aerospace professionals, including engineers, maintenance teams, and operators, with access to an extensive trove of meticulously vetted data sourced from diverse sensors and systems, all securely preserved on the blockchain. The hallmark of this data lies in its transparency and invulnerability to tampering, establishing a robust foundation for making well-informed decisions.

Engineers, fueled by this wealth of accurate historical data, gain the confidence to fine-tune aircraft designs and systems with precision, driving innovation and efficiency in aerospace engineering. Maintenance teams find themselves equipped with a comprehensive maintenance history, enabling them to proactively schedule interventions and optimize the performance of vital components, thereby elevating safety standards. For operators, real-time adjustments to flight plans, rooted in accurate and reliable data, translate into more streamlined operations, reduced fuel consumption, and diminished emissions – crucial factors in the pursuit of sustainability within the aerospace industry.

The outcome of this data-driven decision-making approach is a multifaceted triumph. Safety is fortified, operational excellence is accentuated, and a relentless commitment to cost-effectiveness prevails, a paramount objective in the aerospace sector. In summation, the fusion of AI and blockchain technologies propels the aerospace industry into an era characterized by heightened efficiency, data-driven ingenuity, enhanced safety measures, environmental responsibility, and a steadfast dedication to economic efficiency.

CONCLUSION

The fusion of AI and blockchain technologies in aerospace engineering marks a pivotal development that establishes a robust ecosystem at the intersection of data collection, security, transparency, and predictive analytics. This synergy is the cornerstone of a paradigm shift in the industry, elevating the integrity of aerospace data management to unprecedented heights. The deployment of advanced sensors continuously captures a wealth of real-time data, from critical engine parameters to external environmental factors, all of which are meticulously recorded and securely stored on the blockchain. This blockchain ledger, with its tamper-proof characteristics, not only safeguards the data but also provides a comprehensive history of each data entry, tracing it back to its source, be it a specific aircraft or a component. The potential of this integration extends to predictive analytics, where AI algorithms delve into the wealth of data, identifying patterns, anomalies, and performance trends. These algorithms provide aerospace engineers, maintenance teams, and operators with the foresight needed for informed decision-making. Maintenance schedules are optimized as AI predicts component wear and tear, leading to proactive interventions and heightened safety. Real-time adjustments to flight paths are made, resulting in more fuel-efficient operations and cost savings for airlines. This new era of data-driven decision-making fosters a culture of innovation in aerospace engineering, where the sky is not the limit, but rather the beginning of a dynamic future defined by enhanced efficiency, reliability, and continuous advancements.

REFERENCES

Chuck, M. C., Parker, R., & Thomson, S. (2019). How Blockchain can improve the aviation industry. *Strategy + Business Online Magazine*. https://www.strategy-business.com/article/How-blockchain-can-improve-the-aviationindustry?gko=9e976

Lin, I. C., & Liao, T. C. (2017). A survey of blockchain security issues and challenges. *International Journal of Network Security*, *19*(5), 653–659.

Liu, Yu, Li, Ji, & Leung. (2020). Blockchain and Machine Learning for Communications and Networking Systems. *IEEE Communications Surveys & Tutorials*. . doi:10.1109/COMST.2020.2975911

Lohmer, J., & Lasch, R. (2020). Blockchain in operations management and manufacturing: Potential and barriers. *Computers & Industrial Engineering*, *149*, 106789. doi:10.1016/j.cie.2020.106789

Longo, F., Nicoletti, L., Padovano, A., d'Atri, G., & Forte, M. (2019). Blockchain-enabled supply chain: An experimental study. *Computers & Industrial Engineering*, *136*, 57–69. doi:10.1016/j.cie.2019.07.026

Madhwal, Y., & Panfilov, P. (2017). Industrial case: Blockchain on aircraft's parts supply chain management. In *American Conference on Proceedings of Information Systems and Workshop on Smart Manufacturing (AMCIS), Boston* (pp. 1–6). Academic Press.

Makridakis, S., Polemitis, A., Giaglis, G., & Louca, S. (2018). Blockchain: The next breakthrough in the rapid progress of AI. *Artificial Intelligence-Emerging Trends and Applications*, 197–219.

Malavolta, G., Moreno-Sanchez, P., Schneidewind, C., Kate, A., & Maffei, M. (2019). Anonymous multi-hop locks for blockchain scalability and interoperability. In *Network and Distributed Systems Security (NDSS) Symposium,* San Diego, USA (pp. 24–27). 10.14722/ndss.2019.23330

Mammadzada, K., Iqbal, M., Milani, F., Garc, L., & Matulevicius, R. (2020). Blockchain Oracles: A framework for Blockchain-based applications. In *Springer International Conference on Business Process Management,* Sevilla, Spain (pp. 19–34). 10.1007/978-3-030-58779-6_2

Niculescu-Mizil. (2020). Blockchain in Aerospace Towards Transformational Air Travel. Academic Press.

Rana, T., Shankar, A., Sultan, M. K., Patan, R., & Balusamy, B. (2019). An intelligent approach for UAV and drone privacy security using blockchain methodology. In *IEEE 9th International Conference on Cloud Computing, Data Science & Engineering (Confluence), Noida, India* (pp. 162–167). 10.1109/CONFLUENCE.2019.8776613

Compilation of References

Abdelbar, A. M., Abdelshahid, S., & Wunsch, D. C. I. (2005). Fuzzy PSO: A generalization of particle swarm optimization. *Proceedings of 2005 IEEE International Joint Conference on Neural Networks (IJCNN '05),* 1086–1091. 10.1109/IJCNN.2005.1556004

Abdulrahman, Y., Arnautović, E., Parezanović, V., & Svetinovic, D. (2023). AI and Blockchain Synergy in Aerospace Engineering: An Impact Survey on Operational Efficiency and Technological Challenges. *IEEE Access : Practical Innovations, Open Solutions, 11,* 87790–87804. doi:10.1109/ACCESS.2023.3305325

Abdulrahman, Y., Parezanovic, V., & Svetinovic, D. (2022). AI-blockchain systems in aerospace engineering and management: Review and challenges. *Proc. 30th Telecommun. Forum (TELFOR),* 1–4. 10.1109/TELFOR56187.2022.9983700

Abubakar, A., Kolo, K., & Zhang, Y. (2019). Machine Learning Approaches in Astrophysics: A Review. *Frontiers in Astronomy and Space Sciences, 6,* 53. doi:10.3389/fspas.2019.00053

Aihara, H., Allende Prieto, C., An, D., Anderson, S. F., Aubourg, É., Balbinot, E., Beers, T. C., Berlind, A. A., Bickerton, S. J., Bizyaev, D., Blanton, M. R., Bochanski, J. J., Bolton, A. S., Bovy, J., Brandt, W. N., Brinkmann, J., Brown, P. J., Brownstein, J. R., Busca, N. G., ... Zehavi, I. (2011). The Eighth Data Release of the Sloan Digital Sky Survey: First Data from SDSS-III. *The Astrophysical Journal. Supplement Series, 193*(2), 29. doi:10.1088/0067-0049/193/2/29

Al-Breiki, H., Rehman, M. H. U., Salah, K., & Svetinovic, D. (2020). Trustworthy blockchain oracles: Review, comparison, and open research challenges. *IEEE Access : Practical Innovations, Open Solutions, 8,* 85675–85685. doi:10.1109/ACCESS.2020.2992698

Alofi, A., Bokhari, M. A., Bahsoon, R., & Hendley, R. (2022). Optimizing the Energy Consumption of Blockchain-Based Systems Using Evolutionary Algorithms: A New Problem Formulation. *IEEE Transactions on Sustainable Computing, 7*(4), 910–922. doi:10.1109/TSUSC.2022.3160491

Amari, H., Houda, Z. A. E., Khoukhi, L., & Belguith, L. H. (2023). Trust management in vehicular ad-hoc networks: Extensive survey. *IEEE Access : Practical Innovations, Open Solutions, 11,* 47659–47680. doi:10.1109/ACCESS.2023.3268991

Amit, S. N. K. B., & Aoki, Y. (2017), Disaster detection from aerial imagery with convolutional neural network. *Knowledge Creation and Intelligent Computing (IES-KCIC), 2017 International Electronics Symposium (IEEE),* 239–245 10.1109/KCIC.2017.8228593

Anuar, H. A. B., Plestan, F., Chriette, A., & Kermorgant, O. (2022). Super-Twisting Sliding Mode Control with Adaptive Gain of Quadrotor with Rigid Manipulator. *2022 16th International Workshop on Variable Structure Systems (VSS)*, 53–58. 10.1109/VSS57184.2022.9902104

Audrey, Q., Robin, V., Gilles, W., & Clementz, G. (2020, September 8). *Blockchain in defence: a breakthrough?* https://finabel.org/blockchain-in-defence-a-breakthrough/#:~:text=The%20advent%20of%20blockchain%20technology,the%20technology%20may%20impede%20them

Awad, M. (2010). An unsupervised artificial neural network method for satellite image segmentation. *International Arab Journal of Information Technology, 7*(2), 199 – 205.

Awad, M., Chehdi, K., & Nasri, A. (2007). Multicomponent Image Segmentation Using a Genetic Algorithm and Artificial Neural Network. *IEEE Geoscience and Remote Sensing Letters*, *4*(4), 571–575. doi:10.1109/LGRS.2007.903064

Aydn, S., & Kahraman, C. (2021). Aviation 4.0 Revolution. In *Intelligent and Fuzzy Techniques in Aviation 4.0: Theory and Applications* (pp. 3–19). Springer International Publishing.

Azari, A., Biersteker, J. B., Dewey, R. M., Doran, G., Forsberg, E. J., Harris, C. D. K., Kerner, H. R., Skinner, K. A., Smith, A. W., Amini, R., Cambioni, S., Poian, V. D., Garton, T. M., Himes, M. D., Millholland, S., & Ruhunusiri, S. (2021). Integrating Machine Learning for Planetary Science: Perspectives for the Next Decade. *Bulletin of the AAS*, *53*(4). Advance online publication. doi:10.3847/25c2cfeb.aa328727

Basiri, M. E., Ghasem-Aghaee, N., & Aghdam, M. H. (2013). Efficient ant colony optimization for image feature selection. *Signal Processing*, *93*(1), 1–14. doi:10.1016/j.sigpro.2012.06.019

Bhandari, A. K., Pare, S., Kumar, A., Singh, S. K., & Kumar, A. (2015). Satellite image segmentation based on different objective functions using genetic algorithm: A comparative study. In *2015 International Conference on Digital Signal Processing* (pp. 1245-1250). IEEE. 10.1109/ICDSP.2015.7251972

Breiman, L. (2001). Random forests. *Machine Learning*, *45*(1), 5–32. doi:10.1023/A:1010933404324

Brown, A. G. A., Vallenari, A., Prusti, T., de Bruijne, J. H. J., Babusiaux, C., Bailer-Jones, C. A. L., Biermann, M., Evans, D. W., Eyer, L., Jansen, F., Jordi, C., Klioner, S. A., Lammers, U., Lindegren, L., Luri, X., Mignard, F., Panem, C., Pourbaix, D., Randich, S., ... Zwitter, T. (2018). Gaia Data Release 2: Summary of the contents and survey properties. *Astronomy & Astrophysics*, *616*, A1. doi:10.1051/0004-6361/201833051

Cao, Q. D., & Choe, Y. (2018), Deep Learning Based Damage Detection on Post Hurricane Satellite Imagery. *arXiv preprint arXiv:1807.01688.*

Celik, T. (2010). Change detection in satellite images using a genetic algorithm approach. *IEEE Geoscience and Remote Sensing Letters*, *7*(2), 388–392. doi:10.1109/LGRS.2009.2037024

Chang, S., Wang, Z., Wang, Y., Tang, J., & Jiang, X. (2019, August). Enabling technologies and platforms to aid digitalization of commercial aviation support, maintenance and health management. In *2019 International Conference on Quality, Reliability, Risk, Maintenance, and Safety Engineering (QR2MSE)* (pp. 926-932). IEEE. 10.1109/QR2MSE46217.2019.9021222

Chang, T.-H., & Svetinovic, D. (2020, January). Improving Bitcoin ownership identification using transaction patterns analysis. *IEEE Transactions on Systems, Man, and Cybernetics. Systems*, *50*(1), 9–20. doi:10.1109/TSMC.2018.2867497

Chen, Q., Tao, M., He, X., & Tao, L. (2021). Fuzzy Adaptive Nonsingular Fixed-Time Attitude Tracking Control of Quadrotor UAVs. *IEEE Transactions on Aerospace and Electronic Systems*, *57*(5), 2864–2877. doi:10.1109/TAES.2021.3067610

ChenX.EderM. A.ShihavuddinA. S. M. (2023). A concept for human-cyber-physical systems of future wind turbines towards Industry 5.0. Authorea Preprints.

Chollet, F. (2015). *Keras*. GitHub repository: https://github.com/fchollet/keras

Chuck, M. C., Parker, R., & Thomson, S. (2019). How Blockchain can improve the aviation industry. *Strategy + Business Online Magazine*. https://www.strategy-business.com/article/How-blockchain-can-improve-the-aviationindustry?gko=9e976

Chu, H., Jing, Q., Chang, Z., Shao, Y., Zhang, X., & Mukherjee, M. (2020). Quadrotor Attitude Control via Feedforward All-Coefficient Adaptive Theory. *IEEE Access : Practical Innovations, Open Solutions*, *8*, 116441–116453. doi:10.1109/ACCESS.2020.3002962

Coello, C., Lamont, G., & Veldhuizen, D. (2007). *Evolutionary Algorithms for Solving Multi-Objective Problems*. Springer.

Corea, F. (2019). The convergence of AI and blockchain. In *Applied Artificial Intelligence: Where AI Can Be Used*. Springer., doi:10.1007/978-3-319-77252-3_4

Crosby, M., Pattanayak, P., Verma, S., & Kalyanaraman, V. (2016). Blockchain technology: Beyond bitcoin. *Appl. Innov.*, *2*(6–10), 71.

Dan, W., Yin, L., Yeying, W., Rui, W., Chen, F., Zhongyang, Z., Dongdong, H., & Liang, G. (2020). Research on aerospace data transaction platform and method based on blockchain. *Proc. 2nd Int. Conf. Inf. Technol. Comput. Appl. (ITCA),* 410–413. 10.1109/ITCA52113.2020.00092

de Bruijne, J. H. J. (2012). Astrometric satellite missions. In EAS Publications Series (Vol. 45, pp. 105–116). doi:10.1051/eas/1245003

De Falco, I., Della Cioppa, A., Maisto, D., & Tarantino, E. (2008). Differential Evolution as a viable tool for satellite image registration. *Applied Soft Computing*, *8*(4), 1453-1462. doi:10.1016/j.asoc.2007.10.013

De Falco, I., Maisto, D., Scafuri, U., Tarantino, E., & Della Cioppa, A. (2007). Distributed Differential Evolution for the Registration of Remotely Sensed Images. In *15th EUROMICRO International Conference on Parallel, Distributed and Network-Based Processing (PDP'07)*, Napoli, Italy (pp. 358-362). 10.1109/PDP.2007.36

Deb, K., Pratap, A., Agarwal, S., & Meyarivan, T. (2002). A fast and elitist multiobjective genetic algorithm: NSGA-II. *IEEE Transactions on Evolutionary Computation*, *6*(2), 182–197. doi:10.1109/4235.996017

Demir, Koperski, Lindenbaum, Pang, Huang, Basu, Hughes, Tuia, & Raskar. (2018). *Deepglobe: A challenge to parse the earth through satellite images*. ArXiv e-prints.

Dhal, K. G., Quraishi, M. I., & Das, S. (2015). Parameterized image enhancement technique using modified differential evolution via chaotic Lévy flight. *International Journal of Applied Metaheuristic Computing*, *6*(3), 69–81. doi:10.4018/ijamc.2015070104

Difrancesco, R. M., Meena, P., & Kumar, G. (2023). How blockchain technology improves sustainable supply chain processes: A practical guide. *Operations Management Research : Advancing Practice Through Research*, *16*(2), 620–641. doi:10.1007/s12063-022-00343-y

Dorigo, M. (1992). *Optimization, learning and natural algorithms* [PhD thesis]. Politecnico di Milano, Italy.

Doshi, J. (2018). Residual Inception Skip Network for Binary Segmentation. *Proceedings of the IEEE Conference on Computer Vision and Pattern Recognition Workshops*, 216–219. 10.1109/CVPRW.2018.00037

Druby, A. (2019). *Predictive analytics: What and why it matters?* https://www.sas.com/en_in/insights/analytics/predictive-analytics.html

Duarte, D., Nex, F., Kerle, N., & Vosselman, G. (2018). Satellite Image Classification of building damages using airborne and satellite image samples in a deep learning approach. *ISPRS Annals of the Photogrammetry, Remote Sensing and Spatial Information Sciences*, *4*(2), 89–96. doi:10.5194/isprs-annals-IV-2-89-2018

Efthymiou, M., McCarthy, K., Markou, C., & O'Connell, J. F. (2022). An exploratory research on blockchain in aviation: The case of maintenance, repair and overhaul (MRO) organizations. *Sustainability (Basel)*, *14*(5), 2643. doi:10.3390/su14052643

Ehsani, F. (2018). *Blockchain in finance: From buzzword to watchword in 2016*. Available: https://www.coindesk.com/markets/2016/12/20/blockchain-in-financefrom-buzzword-to-watchword-in-2016/

Elangovan, U. (2021). *Industry 5.0: The future of the industrial economy*. CRC Press. doi:10.1201/9781003190677

El-Sousy, F. F. M., Alattas, K. A., Mofid, O., Mobayen, S., & Fekih, A. (2022). Robust Adaptive Super-Twisting Sliding Mode Stability Control of Underactuated Rotational Inverted Pendulum With Experimental Validation. *IEEE Access : Practical Innovations, Open Solutions*, *10*, 100857–100866. doi:10.1109/ACCESS.2022.3208412

El-Sousy, F. F. M., Amin, M. M., & Mohammed, O. A. (2022). Robust Adaptive Neural Network Tracking Control With Optimized Super-Twisting Sliding-Mode Technique for Induction Motor Drive System. *IEEE Transactions on Industry Applications*, *58*(3), 4134–4157. doi:10.1109/TIA.2022.3160136

Ezhilarasan, K., & Kiruthika, K. (2020). Automatic Building Detection Using Thresholding and Morphological Operations. In *National Conference on Application of Geospatial Technologies in Management of Natural Resources and Disaster (UGC Sponsored) - (AGTMNRD-2020)* (pp. 138-141). Academic Press.

Ezhilarasan, K., & Vennila, M. (2020). An automatic shadow detection using mathematical morphological operations. In *National Conference on Application of Geospatial Technologies in Management of Natural Resources and Disaster (UGC Sponsored) - (AGTMNRD-2020)* (pp. 134-137). Academic Press.

Ezhilarasan, K., Somasundaram, K., Kalaiselvi, T., Praveenkumar, S., Karthigai Selvi, S., & Jeevarekha, A. (2024). Bio-Inspired Algorithms Used in Medical Image Processing. In U. Vignesh, M. Manikandan, & R. Doshi (Eds.), *Bio-Inspired Optimization Techniques in Blockchain Systems* (pp. 1–300). IGI Global. doi:10.4018/979-8-3693-1131-8

Ezzat, D., Hassanien, A. E., Darwish, A., Yahia, M., Ahmed, A., & Abdelghafar, S. (2021). Multi-objective hybrid artificial intelligence approach for fault diagnosis of aerospace systems. *IEEE Access : Practical Innovations, Open Solutions*, *9*, 41717–41730. doi:10.1109/ACCESS.2021.3064976

Ferrer, E. C. (2018). The blockchain: A new framework for robotic swarm systems. In *Springer Proceedings of the Future Technologies Conference.* Springer.

Fujita, A., Sakurada, K., Imaizumi, T., Ito, R., Hikosaka, S., & Nakamura, R. (2017). Damage detection from aerial images via convolutional neural networks. *IAPR International Conference on Machine Vision Applications, Nagoya, Japan*, 8–12. 10.23919/MVA.2017.7986759

Ghadiri, H., Emami, M., & Khodadadi, H. (2021). Adaptive super-twisting non-singular terminal sliding mode control for tracking of quadrotor with bounded disturbances. *Aerospace Science and Technology*, *112*, 106616. doi:10.1016/j.ast.2021.106616

Ghimire, T., Joshi, A., Sen, S., Kapruan, C., Chadha, U., & Selvaraj, S. K. (2022, January). Blockchain in additive manufacturing processes: Recent trends & its future possibilities. *Materials Today: Proceedings*, *50*, 2170–2180. doi:10.1016/j.matpr.2021.09.444

Ghobakhloo, M., Iranmanesh, M., Morales, M. E., Nilashi, M., & Amran, A. (2023). Actions and approaches for enabling Industry 5.0-driven sustainable industrial transformation: A strategy roadmap. *Corporate Social Responsibility and Environmental Management, 30*(3), 1473–1494. doi:10.1002/csr.2431

Gunasekara, H. G., Sridarran, P., & Rajaratnam, D. (2022, May). Effective use of blockchain technology for facilities management procurement process. *Journal of Facilities Management, 20*(3), 452–468. doi:10.1108/JFM-10-2020-0077

Gupta, R., Kumari, A., & Tanwar, S. (2021). Fusion of blockchain and artificial intelligence for secure drone networking underlying 5G communications. *Transactions on Emerging Telecommunications Technologies, 32*(1), e4176. doi:10.1002/ett.4176

Hackius, N., & Petersen, M. (2017). Blockchain in logistics and supply chain: Trick or treat? In *Proceedings of the Hamburg International Conference of Logistics (HICL),* Hamburg University of Technology, Germany. epubli GmbH.

Hartley, C., & Conway, S. (2018, June 12). *Vast Majority of Aerospace and Defense Companies Plan to Integrate Blockchain by 2021, Accenture Report Finds.* https://newsroom.accenture.com/news/vast-majority-of-aerospace-and-defense-companies-plan-to-integrate-blockchain-by-2021-accenture-report-finds.htm

He, K., Zhang, X., Ren, S., & Sun, J. (2015). Delving deep into rectifiers: Surpassing human-level performance on imagenet classification. *Proceedings of the IEEE international conference on computer vision*, 1026–1034. 10.1109/ICCV.2015.123

He, K., Zhang, X., Ren, S., & Sun, J. (2016). Deep residual learning for image recognition. In *Proceedings of the IEEE conference on computer vision and pattern recognition (CVPR)* (pp. 770-778). IEEE.

Henderson, A., Harbour, S., & Cohen, K. (2022). Toward airworthiness certification for artificial intelligence (AI) in aerospace systems. *Proc. IEEE/AIAA 41st Digit. Avionics Syst. Conf. (DASC),* 1–10. 10.1109/DASC55683.2022.9925740

Hepner, G., & Ritter, N. (1989). *Application of an artificial neural network to land cover classification of thematic mapper imagery.* JPL Internal Technical Report.

Hoseini, P., & Shayesteh, M. G. (2013). Efficient contrast enhancement of images using hybrid ant colony optimisation, genetic algorithm, and simulated annealing. *Digital Signal Processing, 23*(3), 879–893. doi:10.1016/j.dsp.2012.12.011

Huang, T., Li, B., Shah, A., Qin, N., & Huang, D. (2019). Fuzzy Sliding Mode Control for a Quadrotor UAV. *2019 IEEE 8th Data Driven Control and Learning Systems Conference (DDCLS),* 672–677. 10.1109/DDCLS.2019.8908982

Huynh-The, T., Gadekallu, T. R., Wang, W., Yenduri, G., Ranaweera, P., Pham, Q.-V., da Costa, D. B., & Liyanage, M. (2023, June). Blockchain for the metaverse: A review. *Future Generation Computer Systems, 143*, 401–419. doi:10.1016/j.future.2023.02.008

Ignatiev, V., Trekin, A., Lobachev, V., Potapov, G., & Burnaev, E. (2018). Targeted change detection in remote sensing images. *arXiv preprint arXiv:1803.05482.*

Jafari, N., Azarian, M., & Yu, H. (2022). Moving from Industry 4.0 to Industry 5.0: What are the implications for smart logistics? *Logistics*, *6*(2), 26. doi:10.3390/logistics6020026

Jain, D. K., Li, Y., Er, M. J., Xin, Q., Gupta, D., & Shankar, K. (2021). Enabling unmanned aerial vehicle borne secure communication with classification framework for industry 5.0. *IEEE Transactions on Industrial Informatics*, *18*(8), 5477–5484. doi:10.1109/TII.2021.3125732

James, T., & Roper, S. (2018). *Launching from earth: The science behind space law and technological developments. In Deep Space Commodities*. Palgrave Macmillan. doi:10.1007/978-3-319-90303-3_3

Jasim, W., & Gu, D. (2014). H∞ for quadrotor attitude stabilization. *2014 UKACC International Conference on Control (CONTROL)*, 19–24. 10.1109/CONTROL.2014.6915109

Javaid, M., & Haleem, A. (2020). Critical components of Industry 5.0 towards a successful adoption in the field of manufacturing. *Journal of Industrial Integration and Management*, *5*(03), 327–348. doi:10.1142/S2424862220500141

Jiang, H., Li, C., Oliva, D., & Peng, X. (2019). Hybrid grasshopper optimization algorithm and differential evolution for multilevel satellite image segmentation. *Remote Sensing (Basel)*, *11*(9), 1134. doi:10.3390/rs11091134

Jing, Q., Chang, Z., Chu, H., Shao, Y., & Zhang, X. (2019). Quadrotor Attitude Control Based on Fuzzy Sliding Mode Control Theory. *2019 Chinese Control Conference (CCC)*, 8360–8364. 10.23919/ChiCC.2019.8865754

Joseph, F. (2018). *How blockchain will simplify and transform the supply chain*. Accenture White Paper.

Kavasidis, I., Lallas, E., Mountzouris, G., Gerogiannis, V. C., & Karageorgos, A. (2023). A federated learning framework for enforcing traceability in manufacturing processes. *IEEE Access : Practical Innovations, Open Solutions*, *11*, 57585–57597. doi:10.1109/ACCESS.2023.3282316

Kavya, J., Prasad, G., & Bharanidharan, N. (2022). Artificial Intelligence, Machine Learning, and Internet of Drones in Medical Applications. In *Bio-Inspired Algorithms and Devices for Treatment of Cognitive Diseases Using Future Technologies* (pp. 180–188). IGI Global. doi:10.4018/978-1-7998-9534-3.ch011

Khunaifi, A., Wijananto, F., Handoyo, K. M., Juliansyah, M. R., Imdad, M. T., Wibowo, M., ... Arief, M. (2023, February). The Utilization of Blockchain Technology in The Airport Industry: A Review. In *Proceeding of International Conference on Information Science and Technology Innovation (ICoSTEC)* (Vol. 2, No. 1, pp. 145-151). 10.35842/icostec.v2i1.53

Kingma, D. P., & Ba, J. (2014). Adam: A method for stochastic optimization. *arXiv preprint arXiv:1412.6980.*

Kumar Nimbark, H. (2023). Development of Algorithm for Blockchain and Artificial Intelligence Collaboration to Identify Travel Path of UAVs. *Tuijin Jishu/Journal of Propulsion Technology, 44*(3), 660-666.

Kumar, G. R., Rencher, R., Fabre, C., Budeanu, D., Markou, C., Jones, K., ... Veluri, S. (2022). Best Practices in Establishing Business Case for Implementing Blockchain Solution in Aerospace. *SAE International Journal of Advances and Current Practices in Mobility, 4*, 1104-1114.

LeCun, Y., Bengio, Y., & Hinton, G. (2015). Deep learning. *Nature, 521*(7553), 436–444. doi:10.1038/nature14539 PMID:26017442

Lee, C. K., Huo, Y. Z., Zhang, S. Z., & Ng, K. K. (2020). Design of a smart manufacturing system with the application of multi-access edge computing and blockchain technology. *IEEE Access : Practical Innovations, Open Solutions, 8*, 28659–28667. doi:10.1109/ACCESS.2020.2972284

Lépine, S., & Shara, M. M. (2005). A Catalog of Northern Stars with Annual Proper Motions Larger than 0.15" (LSPM-NORTH Catalog). *The Astronomical Journal, 129*(3), 1483–1522. doi:10.1086/427854

Lian, S., Meng, W., Lin, Z., Shao, K., Zheng, J., Li, H., & Lu, R. (2022). Adaptive Attitude Control of a Quadrotor Using Fast Nonsingular Terminal Sliding Mode. *IEEE Transactions on Industrial Electronics, 69*(2), 1597–1607. doi:10.1109/TIE.2021.3057015

Lin, I. C., & Liao, T. C. (2017). A survey of blockchain security issues and challenges. *International Journal of Network Security, 19*(5), 653–659.

Liu, Yu, Li, Ji, & Leung. (2020). Blockchain and Machine Learning for Communications and Networking Systems. *IEEE Communications Surveys & Tutorials.* . doi:10.1109/COMST.2020.2975911

Liu, X., Liu, L., & Zhang, J. (2008). Ant colony optimization for classification of remote-sensing imagery. *IEEE Transactions on Geoscience and Remote Sensing, 46*(12), 4190–4199. doi:10.1109/TGRS.2008.2001754

Li, W., Su, Z., Li, R., Zhang, K., & Wang, Y. (2020). Blockchain-based data security for artificial intelligence applications in 6G networks. *IEEE Network, 34*(6), 31–37. doi:10.1109/MNET.021.1900629

Li, X., Jiang, P., Chen, T., Luo, X., & Wen, Q. (2020). A survey on the security of blockchain systems. *Future Generation Computer Systems, 107*, 841–853. doi:10.1016/j.future.2017.08.020

Li, Z., Zhai, J., & Karimi, H. R. (2021). Adaptive finite-time super-twisting sliding mode control for robotic manipulators with control backlash. *International Journal of Robust and Nonlinear Control, 31*(17), 8537–8550. doi:10.1002/rnc.5744

Lohmer, J., & Lasch, R. (2020). Blockchain in operations management and manufacturing: Potential and barriers. *Computers & Industrial Engineering, 149*, 106789. doi:10.1016/j.cie.2020.106789

Longo, F., Nicoletti, L., Padovano, A., d'Atri, G., & Forte, M. (2019). Blockchain-enabled supply chain: An experimental study. *Computers & Industrial Engineering*, *136*, 57–69. doi:10.1016/j.cie.2019.07.026

Lv, Z., Chen, D., Lou, R., & Alazab, A. (2021, April). Artificial intelligence for securing industrial-based cyber–physical systems. *Future Generation Computer Systems*, *117*, 291–298. doi:10.1016/j.future.2020.12.001

Madhwal, Y., & Panfilov, P. (2017). Industrial case: Blockchain on aircraft's parts supply chain management. In *American Conference on Proceedings of Information Systems and Workshop on Smart Manufacturing (AMCIS), Boston* (pp. 1–6). Academic Press.

Makridakis, S., Polemitis, A., Giaglis, G., & Louca, S. (2018). Blockchain: The next breakthrough in the rapid progress of AI. *Artificial Intelligence-Emerging Trends and Applications*, 197–219.

Malavolta, G., Moreno-Sanchez, P., Schneidewind, C., Kate, A., & Maffei, M. (2019). Anonymous multi-hop locks for blockchain scalability and interoperability. In *Network and Distributed Systems Security (NDSS) Symposium,* San Diego, USA (pp. 24–27). 10.14722/ndss.2019.23330

Mammadzada, K., Iqbal, M., Milani, F., Garc, L., & Matulevicius, R. (2020). Blockchain Oracles: A framework for Blockchain-based applications. In *Springer International Conference on Business Process Management,* Sevilla, Spain (pp. 19–34). 10.1007/978-3-030-58779-6_2

Mandolla, C., Petruzzelli, A. M., Percoco, G., & Urbinati, A. (2019, August). Building a digital twin for additive manufacturing through the exploitation of blockchain: A case analysis of the aircraft industry. *Computers in Industry*, *109*, 134–152. doi:10.1016/j.compind.2019.04.011

Manoharan, H., Teekaraman, Y., Kuppusamy, R., Kaliyan, N., & Thelkar, A. R. (2022). Examining the effect of cyber twin and blockchain technologies for industrial applications using AI. *Mathematical Problems in Engineering*, *2022*, 2022. doi:10.1155/2022/3048038

Martin, A.-S., & Freeland, S. (2021). The advent of artificial intelligence in space activities: New legal challenges. *Space Policy*, *55*(Feb), 101408. doi:10.1016/j.spacepol.2020.101408

Martintoni, D., Senni, V., Marin, E. G., & Gutierrez, A. J. C. (2022). Sensitive information protection in blockchain-based supply-chain management for aerospace. *Proc. IEEE Int. Conf. Omni-layer Intell. Syst. (COINS),* 1–8. 10.1109/COINS54846.2022.9854974

Massaro, A. (2021). *Electronics in Advanced Research Industries: Industry 4.0 to Industry 5.0 Advances.* John Wiley & Sons. doi:10.1002/9781119716907

Mattioli, J., Perico, P., & Robic, P.-O. (2020). Improve total production maintenance with artificial intelligence. *Proc. 3rd Int. Conf. Artif. Intell. Ind. (AI4I),* 56–59. 10.1109/AI4I49448.2020.00019

Maulik, U., & Saha, I. (2009). Modified differential evolution based fuzzy clustering for pixel classification in remote sensing imagery. *Pattern Recognition*, *42*(9), 2135–2149. doi:10.1016/j.patcog.2009.01.011

McCulloch, W. S., & Pitts, W. (1943). A logical calculus of the ideas immanent in nervous activity. *The Bulletin of Mathematical Biophysics*, *5*(4), 115–133. doi:10.1007/BF02478259

Mehrotra, A., Singh, K. K., & Singh, P. K. (2014). Unsupervised change detection in satellite images using ant colony optimization clustering. In *2014 International Conference on Computing for Sustainable Global Development (INDIACom)* (pp. 410-415). IEEE. 10.1109/IndiaCom.2014.6828169

Mishra, S., & Singh, S. (2017). Satellite image segmentation based on differential evolution. *Proceedings of the International Conference on Intelligent Sustainable Systems (ICISS 2017),* 623. https://doi.org/10.1109/ICISS.2017.8388766

Modi, A., Joshi, N., & Mehta, A. (2023). Robust discrete-time super twisting formation protocol for a 6-DOF Quadcopter swarm. *ISA Transactions*, 177–187. Advance online publication. doi:10.1016/j.isatra.2023.09.029 PMID:37798204

Mondragon, A. E. C., Mondragon, C. E. C., & Coronado, E. S. (2018). Exploring the applicability of blockchain technology to enhance manufacturing supply chains in the composite materials industry. *Proc. IEEE Int. Conf. Appl. Syst. Invention (ICASI),* 1300–1303. 10.1109/ICASI.2018.8394531

Nakamura, M., Takaya, K., Ohta, H., Shibayama, K., & Kroumov, V. (2019). Quadrotor Modeling and Simulation for Industrial Application. *2019 23rd International Conference on System Theory, Control and Computing (ICSTCC)*, 37–42. 10.1109/ICSTCC.2019.8885708

Nartey, C., Tchao, E. T., Gadze, J. D., Yeboah-Akowuah, B., Nunoo-Mensah, H., Welte, D., & Sikora, A. (2022). Blockchain-IoT peer device storage optimization using an advanced time-variant multi-objective particle swarm optimization algorithm. *EURASIP Journal on Wireless Communications and Networking*, *2022*(1), 1–27. doi:10.1186/s13638-021-02074-3

Nguyen, T. A., Kaliappan, V. K., Jeon, S., Jeon, K. S., Lee, J. W., & Min, D. (2021, November). Blockchain Empowered Federated Learning with Edge Computing for Digital Twin Systems in Urban Air Mobility. In *Asia-Pacific International Symposium on Aerospace Technology* (pp. 935-950). Singapore: Springer Nature Singapore.

Nguyen, T., Katila, R., & Gia, T. N. (2023). An advanced Internet-of-Drones System with Blockchain for improving quality of service of Search and Rescue: A feasibility study. *Future Generation Computer Systems*, *140*, 36–52. doi:10.1016/j.future.2022.10.002

Niculescu-Mizil. (2020). Blockchain in Aerospace Towards Transformational Air Travel. Academic Press.

Obias, K. C. U., Say, M. F. Q., Fernandez, E. A. V., Chua, A. Y., & Sybingco, E. (2019). A Study of the Interaction of Proportional-Integral- Derivative (PID) Control in a Quadcopter Unmanned Aerial Vehicle (UAV) Using Design of Experiment. *2019 IEEE 11th International Conference on Humanoid, Nanotechnology, Information Technology, Communication and Control, Environment, and Management (HNICEM)*, 1–4. 10.1109/HNICEM48295.2019.9072806

Oche, P. A., Ewa, G. A., & Ibekwe, N. (2021). Applications and challenges of artificial intelligence in space missions. *IEEE Access : Practical Innovations, Open Solutions*, 1. doi:10.1109/ACCESS.2021.3132500

Oliva-Palomo, F., Munoz-Vazquez, A. J., Sanchez-Orta, A., Parra-Vega, V., Izaguirre-Espinosa, C., & Castillo, P. (2019). A Fractional Nonlinear PI-Structure Control for Robust Attitude Tracking of Quadrotors. *IEEE Transactions on Aerospace and Electronic Systems*, *55*(6), 2911–2920. doi:10.1109/TAES.2019.2893817

Perryman, M. A. C. (2001). The Hipparcos and Tycho catalogues. Astrometric and photometric star catalogues derived from the ESA Hipparcos Space Astrometry Mission. *Astronomy & Astrophysics*, *369*, 339–363. doi:10.1051/0004-6361:20010085

Poduval, B., Pitman, K. M., Verkhoglyadova, O., & Wintoft, P. (2023). Editorial: Applications of statistical methods and machine learning in the space sciences. *Frontiers in Astronomy and Space Sciences*, *10*, 1163530. doi:10.3389/fspas.2023.1163530

Prasad, G. (2023). Internet of Unmanned Aerial Vehicle (IOU) in Industry 5.0. In Advanced Research and Real-World Applications of Industry 5.0 (pp. 178-188). IGI Global.

Prasad, G., Saravanan, K., Naveen, R., Nagaraj, V., Dinesh, R., & Anirudh, K. (2023). Influence of Machine Learning Technique in Unmanned Aerial Vehicle. In *Internet of Drones* (pp. 43–57). CRC Press. doi:10.1201/9781003252085-3

Qiu, F., & Jensen, J. R. (2003). Building classification models with combined neural networks for high-resolution satellite imagery. *International Journal of Remote Sensing*, *24*(23), 4551–4570. doi:10.1080/0143116031000116414

Rahayu, S. B., Jusoh, N., Halip, M. H. M., Taib, S. M., & Lee, M. G. (2021). A Conceptual Model of Military Blockchain for Repair Parts Supply Chain Management. In *International Conference on Computer & Information Sciences (ICCOINS)*. IEEE. 10.1109/ICCOINS49721.2021.9497227

Rahayu, S. B., Kamarudin, N. D., & Azahari, A. M. (2019). Military blockchain for supply chain management. *Journal of Education and Social Sciences*, *13*(1), 353–361.

Rahman, M. A., & Wang, Y. (2016). Optimizing intersection-over-union in deep neural networks for image segmentation. In *International Symposium on Visual Computing*. Springer. 10.1007/978-3-319-50835-1_22

Raj, P., Nagarajan, G., & Minu, R. I. (Eds.). (2022). *Applied Edge AI: Concepts, Platforms, and Industry Use Cases*. CRC Press.

Rana, T., Shankar, A., Sultan, M. K., Patan, R., & Balusamy, B. (2019). An intelligent approach for UAV and drone privacy security using blockchain methodology. In *IEEE 9th International Conference on Cloud Computing, Data Science & Engineering (Confluence), Noida, India* (pp. 162–167). 10.1109/CONFLUENCE.2019.8776613

Raoofi, T., & Yasar, S. (2023). Analysis of frontier digital technologies in continuing airworthiness management frameworks and applications. *Aircraft Engineering and Aerospace Technology*, *95*(10), 1669–1677. doi:10.1108/AEAT-06-2022-0166

Ratnakumar, R., & Nanda, S. J. (2019). A low complexity hardware architecture of K-means algorithm for real-time satellite image segmentation. *Multimedia Tools and Applications*, *78*(9), 11949–11981. doi:10.1007/s11042-018-6726-6

Ratnakumar, R., & Nanda, S. J. (2021). A high speed roller dung beetles clustering algorithm and its architecture for real-time image segmentation. *Applied Intelligence*, *51*(7), 4682–4713. doi:10.1007/s10489-020-02067-7

Ratnakumar, R., & Nanda, S. J. (2021, January). An improved genetic clustering architecture for real-time satellite image segmentation. In *2021 International Conference on Advances in Technology, Management & Education (ICATME)* (pp. 123-128). IEEE. 10.1109/ICATME50232.2021.9732768

Ratnakumar, R., & Nanda, S. J. (2023). A high-speed unsupervised hardware architecture for rapid diagnosis of COVID-19. *International Journal of Circuit Theory and Applications*, *51*(1), 437–474. doi:10.1002/cta.3417

Reyna, A., Martín, C., Chen, J., Soler, E., & Díaz, M. (2018, November). On blockchain and its integration with IoT. Challenges and opportunities. *Future Generation Computer Systems*, *88*, 173–190. doi:10.1016/j.future.2018.05.046

Sánchez Pérez, A. M., Tarifa Fernández, J., & Cruz Rambaud, S. (2020). Assessing blockchain investments through the learning option: An application to the automotive and aerospace industry. *Mathematics*, *8*(12), 2213. doi:10.3390/math8122213

Sarkar, S., Das, S., & Chaudhuri, S. S. (2015). Hyper-Spectral Image Segmentation using Rényi Entropy based Multi-Level Thresholding Aided with Differential Evolution. *Expert Systems with Applications*, *42*(7), 3606–3618. doi:10.1016/j.eswa.2015.01.035

Sasaki, Y. (2007). The truth of the F-measure. *Teach Tutor mater, 1*(5), 1–5.

Savsani, V., & Tahwid, A. (2017). Non-dominated sorting moth flame optimization (NS-MFO) for multi-objective problems. *Engineering Applications of Artificial Intelligence*, *63*, 20–32. doi:10.1016/j.engappai.2017.04.018

Seeber, R., Horn, M., & Fridman, L. (2018). A Novel Method to Estimate the Reaching Time of the Super-Twisting Algorithm. *IEEE Transactions on Automatic Control*, *63*(12), 4301–4308. doi:10.1109/TAC.2018.2812789

Shao, X., Sun, G., Yao, W., Liu, J., & Wu, L. (2022). Adaptive Sliding Mode Control for Quadrotor UAVs With Input Saturation. *IEEE/ASME Transactions on Mechatronics*, *27*(3), 1498–1509. doi:10.1109/TMECH.2021.3094575

Sharma, R., & Arya, R. (2022). UAV based long range environment monitoring system with Industry 5.0 perspectives for smart city infrastructure. *Computers & Industrial Engineering*, *168*, 108066. doi:10.1016/j.cie.2022.108066

Shehzad, M. F., Bilal, A., & Ahmad, H. (2019). Position & Attitude Control of an Aerial Robot (Quadrotor) With Intelligent PID and State feedback LQR Controller: A Comparative Approach. *2019 16th International Bhurban Conference on Applied Sciences and Technology (IBCAST)*, 340–346. 10.1109/IBCAST.2019.8667170

Shmelova, T., Sikirda, Y., & Sterenharz, A. (Eds.). (2019). *Handbook of Research on Artificial intelligence applications in the aviation and aerospace industries*. IGI Global.

Shmelova, T., Sterenharz, A., & Dolgikh, S. (2020). Artificial intelligence in aviation industries: methodologies, education, applications, and opportunities. In *Handbook of research on artificial intelligence applications in the aviation and aerospace industries* (pp. 1–35). IGI Global. doi:10.4018/978-1-7998-1415-3.ch001

Simonyan, K., & Zisserman, A. (2014). Very deep convolutional networks for large-scale image recognition. *arXiv preprint arXiv:1409.1556.*

Singh, N., & Vardhan, M. (2020). Computing Optimal Block Size for Blockchain based Applications with Contradictory Objectives. *Procedia Computer Science*, *171*, 1389–1398. doi:10.1016/j.procs.2020.04.149

Singh, P., Elmi, Z., Lau, Y. Y., Borowska-Stefańska, M., Wiśniewski, S., & Dulebenets, M. A. (2022). Blockchain and AI technology convergence: Applications in transportation systems. *Vehicular Communications*, *38*, 100521. doi:10.1016/j.vehcom.2022.100521

Srivastava, N., Hinton, G., Krizhevsky, A., Sutskever, I., & Salakhutdinov, R. (2014). Dropout: A simple way to prevent neural networks from overfitting. *Journal of Machine Learning Research*, *15*(1), 1929–1958.

Storn, R., & Price, K. (1997). Differential evolution - a simple and efficient heuristic for global optimization over continuous spaces. *Journal of Global Optimization*, *11*(4), 341–359. doi:10.1023/A:1008202821328

Subburaj, B., Jayachandran, U. M., Arumugham, V., & Suthanthira Amalraj, M. J. A. (2023). A Self-Adaptive Trajectory Optimization Algorithm Using Fuzzy Logic for Mobile Edge Computing System Assisted by Unmanned Aerial Vehicle. *Drones (Basel)*, *7*(4), 1–19. doi:10.3390/drones7040266

Subburaj, B., & Suthanthira Amalraj, M. J. A. (2021). A robust and adaptive fuzzy logic based differential evolution algorithm using population diversity tuning for multi-objective optimization. *Engineering Applications of Artificial Intelligence*, *102*, 104240. doi:10.1016/j.engappai.2021.104240

Sudhan, A., & Nene, M. J. (2017). Employability of Blockchain Technology in Defence Applications. In *Proceedings of the International Conference on Intelligent Sustainable Systems (ICISS 2017)*. IEEE. 10.1109/ISS1.2017.8389247

Sunny, J., Undralla, N., & Pillai, V. M. (2020). Supply chain transparency through blockchain-based traceability: An overview with demonstration. *Computers & Industrial Engineering*, *150*, 106895. doi:10.1016/j.cie.2020.106895

Tang, K. S., Man, K. F., Kwong, S., & He, Q. (1996). Genetic algorithms and their applications. *IEEE Signal Processing Magazine*, *13*(6), 22–37. doi:10.1109/79.543973

Thakur, P., & Sehgal, V. K. (2021). Emerging architecture for heterogeneous smart cyber-physical systems for industry 5.0. *Computers & Industrial Engineering*, *162*, 107750. doi:10.1016/j.cie.2021.107750

Ting, Y. S., & Ranganath, S. (2019). Machine Learning for Astronomy. *Annual Review of Statistics and Its Application*, *6*, 399–426. doi:10.1146/annurev-statistics-030718-105348

Tripathi, V. K., Kamath, A. K., Behera, L., Verma, N. K., & Nahavandi, S. (2022). An Adaptive Fast Terminal Sliding-Mode Controller With Power Rate Proportional Reaching Law for Quadrotor Position and Altitude Tracking. *IEEE Transactions on Systems, Man, and Cybernetics. Systems*, *52*(6), 3612–3625. doi:10.1109/TSMC.2021.3072099

Tso, B., & Mather, P. M. (1999). Classification of multisource remote sensing data using a genetic algorithm and Markov random fields. *IEEE Transactions on Geoscience and Remote Sensing*, *37*(3), 1367–1377. doi:10.1109/36.763284

Ullah, N., Mehmood, Y., Aslam, J., Ali, A., & Iqbal, J. (2021). UAVs-UGV Leader Follower Formation Using Adaptive Non-Singular Terminal Super Twisting Sliding Mode Control. *IEEE Access : Practical Innovations, Open Solutions*, *9*, 74385–74405. doi:10.1109/ACCESS.2021.3081483

Vaiyapuri, T., Shankar, K., Rajendran, S., Kumar, S., Acharya, S., & Kim, H. (2023). Blockchain assisted data edge verification with consensus algorithm for machine learning assisted IoT. *IEEE Access : Practical Innovations, Open Solutions*, *11*, 55370–55379. doi:10.1109/ACCESS.2023.3280798

Voigt, S., Giulio-Tonolo, F., Lyons, J., Kucera, J., Jones, B., Schneiderhan, T., Platzeck, G., Kaku, K., Hazarika, M. K., Czaran, L., Li, S., Pedersen, W., James, G. K., Proy, C., Muthike, D. M., Bequignon, J., & Guha-Sapir, D. (2016). Global trends in satellite-based emergency mapping. *Science*, *353*(6296), 247–252. doi:10.1126/science.aad8728 PMID:27418503

Walthall, R. (2020). *Unsettled Topics Concerning Adopting Blockchain Technology in Aerospace* (No. EPR2020021). SAE Technical Paper.

Wasim Ahmad, R., Hasan, H., Yaqoob, I., Salah, K., Jayaraman, R., & Omar, M. (2021). Blockchain for aerospace and defense: Opportunities and open research challenges. *Computers & Industrial Engineering*, *151*, 1–14. doi:10.1016/j.cie.2020.106982

Xu, B., & Lu, X. (2020). An Online Adaptive Control Strategy for Trajectory Tracking of Quadrotors Based on Fuzzy Approximation and Robust Sliding Mode Algorithm. *IEEE Access : Practical Innovations, Open Solutions*, *8*, 215327–215342. doi:10.1109/ACCESS.2020.3039546

Yadav, J. K., Verma, D. C., Jangirala, S., Srivastava, S. K., & Aman, M. N. (2022). Blockchain for aviation industry: Applications and used cases. In *ICT Analysis and Applications* (pp. 475–486). Springer Singapore. doi:10.1007/978-981-16-5655-2_46

Yogi, S. C., Tripathi, V. K., & Behera, L. (2021). Adaptive Integral Sliding Mode Control Using Fully Connected Recurrent Neural Network for Position and Attitude Control of Quadrotor. *IEEE Transactions on Neural Networks and Learning Systems*, *32*(12), 5595–5609. doi:10.1109/TNNLS.2021.3071020 PMID:33881998

Zavareh, B., Foroozan, H., Gheisarnejad, M., & Khooban, M. H. (2021). New trends on digital twin-based blockchain technology in zero-emission ship applications. *Naval Engineers Journal*, *133*(3), 115–135.

Zhai, J., & Li, Z. (2022). Fast-Exponential Sliding Mode Control of Robotic Manipulator With Super-Twisting Method. *IEEE Transactions on Circuits and Wystems. II, Express Briefs*, *69*(2), 489–493. doi:10.1109/TCSII.2021.3081147

Zhang, C., Wang, Z., Zhou, G., Chang, F., Ma, D., Jing, Y., Cheng, W., Ding, K., & Zhao, D. (2023). Towards new-generation human-centric smart manufacturing in Industry 5.0: A systematic review. *Advanced Engineering Informatics*, *57*, 102121. doi:10.1016/j.aei.2023.102121

Zhang, Q., & Li, H. (2007). MOEA/D: A multiobjective evolutionary algorithm based on decomposition. *IEEE Transactions on Evolutionary Computation*, *11*(6), 712–731. doi:10.1109/TEVC.2007.892759

Related References

To continue our tradition of advancing information science and technology research, we have compiled a list of recommended IGI Global readings. These references will provide additional information and guidance to further enrich your knowledge and assist you with your own research and future publications.

Aasi, P., Rusu, L., & Vieru, D. (2017). The Role of Culture in IT Governance Five Focus Areas: A Literature Review. *International Journal of IT/Business Alignment and Governance, 8*(2), 42-61. https://doi.org/ doi:10.4018/IJITBAG.2017070103

Abdrabo, A. A. (2018). Egypt's Knowledge-Based Development: Opportunities, Challenges, and Future Possibilities. In A. Alraouf (Ed.), *Knowledge-Based Urban Development in the Middle East* (pp. 80–101). Hershey, PA: IGI Global. doi:10.4018/978-1-5225-3734-2.ch005

Abu Doush, I., & Alhami, I. (2018). Evaluating the Accessibility of Computer Laboratories, Libraries, and Websites in Jordanian Universities and Colleges. *International Journal of Information Systems and Social Change*, *9*(2), 44–60. doi:10.4018/IJISSC.2018040104

Adegbore, A. M., Quadri, M. O., & Oyewo, O. R. (2018). A Theoretical Approach to the Adoption of Electronic Resource Management Systems (ERMS) in Nigerian University Libraries. In A. Tella & T. Kwanya (Eds.), *Handbook of Research on Managing Intellectual Property in Digital Libraries* (pp. 292–311). Hershey, PA: IGI Global. doi:10.4018/978-1-5225-3093-0.ch015

Afolabi, O. A. (2018). Myths and Challenges of Building an Effective Digital Library in Developing Nations: An African Perspective. In A. Tella & T. Kwanya (Eds.), *Handbook of Research on Managing Intellectual Property in Digital Libraries* (pp. 51–79). Hershey, PA: IGI Global. doi:10.4018/978-1-5225-3093-0.ch004

Agarwal, P., Kurian, R., & Gupta, R. K. (2022). Additive Manufacturing Feature Taxonomy and Placement of Parts in AM Enclosure. In S. Salunkhe, H. Hussein, & J. Davim (Eds.), *Applications of Artificial Intelligence in Additive Manufacturing* (pp. 138–176). IGI Global. https://doi.org/10.4018/978-1-7998-8516-0.ch007

Al-Alawi, A. I., Al-Hammam, A. H., Al-Alawi, S. S., & AlAlawi, E. I. (2021). The Adoption of E-Wallets: Current Trends and Future Outlook. In Y. Albastaki, A. Razzaque, & A. Sarea (Eds.), *Innovative Strategies for Implementing FinTech in Banking* (pp. 242–262). IGI Global. https://doi.org/10.4018/978-1-7998-3257-7.ch015

Alsharo, M. (2017). Attitudes Towards Cloud Computing Adoption in Emerging Economies. *International Journal of Cloud Applications and Computing*, *7*(3), 44–58. doi:10.4018/IJCAC.2017070102

Amer, T. S., & Johnson, T. L. (2017). Information Technology Progress Indicators: Research Employing Psychological Frameworks. In A. Mesquita (Ed.), *Research Paradigms and Contemporary Perspectives on Human-Technology Interaction* (pp. 168–186). Hershey, PA: IGI Global. doi:10.4018/978-1-5225-1868-6.ch008

Andreeva, A., & Yolova, G. (2021). Liability in Labor Legislation: New Challenges Related to the Use of Artificial Intelligence. In B. Vassileva & M. Zwilling (Eds.), *Responsible AI and Ethical Issues for Businesses and Governments* (pp. 214–232). IGI Global. https://doi.org/10.4018/978-1-7998-4285-9.ch012

Anohah, E. (2017). Paradigm and Architecture of Computing Augmented Learning Management System for Computer Science Education. *International Journal of Online Pedagogy and Course Design*, *7*(2), 60–70. doi:10.4018/IJOPCD.2017040105

Anohah, E., & Suhonen, J. (2017). Trends of Mobile Learning in Computing Education from 2006 to 2014: A Systematic Review of Research Publications. *International Journal of Mobile and Blended Learning*, *9*(1), 16–33. doi:10.4018/IJMBL.2017010102

Arbaiza, C. S., Huerta, H. V., & Rodriguez, C. R. (2021). Contributions to the Technological Adoption Model for the Peruvian Agro-Export Sector. *International Journal of E-Adoption*, *13*(1), 1–17. https://doi.org/10.4018/IJEA.2021010101

Bailey, E. K. (2017). Applying Learning Theories to Computer Technology Supported Instruction. In M. Grassetti & S. Brookby (Eds.), *Advancing Next-Generation Teacher Education through Digital Tools and Applications* (pp. 61–81). Hershey, PA: IGI Global. doi:10.4018/978-1-5225-0965-3.ch004

Baker, J. D. (2021). Introduction to Machine Learning as a New Methodological Framework for Performance Assessment. In M. Bocarnea, B. Winston, & D. Dean (Eds.), *Handbook of Research on Advancements in Organizational Data Collection and Measurements: Strategies for Addressing Attitudes, Beliefs, and Behaviors* (pp. 326–342). IGI Global. https://doi.org/10.4018/978-1-7998-7665-6.ch021

Banerjee, S., Sing, T. Y., Chowdhury, A. R., & Anwar, H. (2018). Let's Go Green: Towards a Taxonomy of Green Computing Enablers for Business Sustainability. In M. Khosrow-Pour (Ed.), *Green Computing Strategies for Competitive Advantage and Business Sustainability* (pp. 89–109). Hershey, PA: IGI Global. doi:10.4018/978-1-5225-5017-4.ch005

Basham, R. (2018). Information Science and Technology in Crisis Response and Management. In M. Khosrow-Pour, D.B.A. (Ed.), Encyclopedia of Information Science and Technology, Fourth Edition (pp. 1407-1418). Hershey, PA: IGI Global. doi:10.4018/978-1-5225-2255-3.ch121

Batyashe, T., & Iyamu, T. (2018). Architectural Framework for the Implementation of Information Technology Governance in Organisations. In M. Khosrow-Pour, D.B.A. (Ed.), Encyclopedia of Information Science and Technology, Fourth Edition (pp. 810-819). Hershey, PA: IGI Global. doi:10.4018/978-1-5225-2255-3.ch070

Bekleyen, N., & Çelik, S. (2017). Attitudes of Adult EFL Learners towards Preparing for a Language Test via CALL. In D. Tafazoli & M. Romero (Eds.), *Multiculturalism and Technology-Enhanced Language Learning* (pp. 214–229). Hershey, PA: IGI Global. doi:10.4018/978-1-5225-1882-2.ch013

Bergeron, F., Croteau, A., Uwizeyemungu, S., & Raymond, L. (2017). A Framework for Research on Information Technology Governance in SMEs. In S. De Haes & W. Van Grembergen (Eds.), *Strategic IT Governance and Alignment in Business Settings* (pp. 53–81). Hershey, PA: IGI Global. doi:10.4018/978-1-5225-0861-8.ch003

Bhardwaj, M., Shukla, N., & Sharma, A. (2021). Improvement and Reduction of Clustering Overhead in Mobile Ad Hoc Network With Optimum Stable Bunching Algorithm. In S. Kumar, M. Trivedi, P. Ranjan, & A. Punhani (Eds.), *Evolution of Software-Defined Networking Foundations for IoT and 5G Mobile Networks* (pp. 139–158). IGI Global. https://doi.org/10.4018/978-1-7998-4685-7.ch008

Bhatt, G. D., Wang, Z., & Rodger, J. A. (2017). Information Systems Capabilities and Their Effects on Competitive Advantages: A Study of Chinese Companies. *Information Resources Management Journal, 30*(3), 41–57. doi:10.4018/IRMJ.2017070103

Bhattacharya, A. (2021). Blockchain, Cybersecurity, and Industry 4.0. In A. Tyagi, G. Rekha, & N. Sreenath (Eds.), *Opportunities and Challenges for Blockchain Technology in Autonomous Vehicles* (pp. 210–244). IGI Global. https://doi.org/10.4018/978-1-7998-3295-9.ch013

Bhyan, P., Shrivastava, B., & Kumar, N. (2022). Requisite Sustainable Development Contemplating Buildings: Economic and Environmental Sustainability. In A. Hussain, K. Tiwari, & A. Gupta (Eds.), *Addressing Environmental Challenges Through Spatial Planning* (pp. 269–288). IGI Global. https://doi.org/10.4018/978-1-7998-8331-9.ch014

Boido, C., Davico, P., & Spallone, R. (2021). Digital Tools Aimed to Represent Urban Survey. In M. Khosrow-Pour D.B.A. (Ed.), *Encyclopedia of Information Science and Technology, Fifth Edition* (pp. 1181-1195). IGI Global. https://doi.org/10.4018/978-1-7998-3479-3.ch082

Borkar, P. S., Chanana, P. U., Atwal, S. K., Londe, T. G., & Dalal, Y. D. (2021). The Replacement of HMI (Human-Machine Interface) in Industry Using Single Interface Through IoT. In R. Raut & A. Mihovska (Eds.), *Examining the Impact of Deep Learning and IoT on Multi-Industry Applications* (pp. 195–208). IGI Global. https://doi.org/10.4018/978-1-7998-7511-6.ch011

Brahmane, A. V., & Krishna, C. B. (2021). Rider Chaotic Biography Optimization-driven Deep Stacked Auto-encoder for Big Data Classification Using Spark Architecture: Rider Chaotic Biography Optimization. *International Journal of Web Services Research, 18*(3), 42–62. https://doi.org/10.4018/ijwsr.2021070103

Burcoff, A., & Shamir, L. (2017). Computer Analysis of Pablo Picasso's Artistic Style. *International Journal of Art, Culture and Design Technologies, 6*(1), 1–18. doi:10.4018/IJACDT.2017010101

Byker, E. J. (2017). I Play I Learn: Introducing Technological Play Theory. In C. Martin & D. Polly (Eds.), *Handbook of Research on Teacher Education and Professional Development* (pp. 297–306). Hershey, PA: IGI Global. doi:10.4018/978-1-5225-1067-3.ch016

Calongne, C. M., Stricker, A. G., Truman, B., & Arenas, F. J. (2017). Cognitive Apprenticeship and Computer Science Education in Cyberspace: Reimagining the Past. In A. Stricker, C. Calongne, B. Truman, & F. Arenas (Eds.), *Integrating an Awareness of Selfhood and Society into Virtual Learning* (pp. 180–197). Hershey, PA: IGI Global. doi:10.4018/978-1-5225-2182-2.ch013

Carneiro, A. D. (2017). Defending Information Networks in Cyberspace: Some Notes on Security Needs. In M. Dawson, D. Kisku, P. Gupta, J. Sing, & W. Li (Eds.), Developing Next-Generation Countermeasures for Homeland Security Threat Prevention (pp. 354-375). Hershey, PA: IGI Global. https://doi.org/ doi:10.4018/978-1-5225-0703-1.ch016

Carvalho, W. F., & Zarate, L. (2021). Causal Feature Selection. In A. Azevedo & M. Santos (Eds.), *Integration Challenges for Analytics, Business Intelligence, and Data Mining* (pp. 145-160). IGI Global. https://doi.org/10.4018/978-1-7998-5781-5.ch007

Chase, J. P., & Yan, Z. (2017). Affect in Statistics Cognition. In *Assessing and Measuring Statistics Cognition in Higher Education Online Environments: Emerging Research and Opportunities* (pp. 144–187). Hershey, PA: IGI Global. doi:10.4018/978-1-5225-2420-5.ch005

Chatterjee, A., Roy, S., & Shrivastava, R. (2021). A Machine Learning Approach to Prevent Cancer. In G. Rani & P. Tiwari (Eds.), *Handbook of Research on Disease Prediction Through Data Analytics and Machine Learning* (pp. 112–141). IGI Global. https://doi.org/10.4018/978-1-7998-2742-9.ch007

Cifci, M. A. (2021). Optimizing WSNs for CPS Using Machine Learning Techniques. In A. Luhach & A. Elçi (Eds.), *Artificial Intelligence Paradigms for Smart Cyber-Physical Systems* (pp. 204–228). IGI Global. https://doi.org/10.4018/978-1-7998-5101-1.ch010

Cimermanova, I. (2017). Computer-Assisted Learning in Slovakia. In D. Tafazoli & M. Romero (Eds.), *Multiculturalism and Technology-Enhanced Language Learning* (pp. 252–270). Hershey, PA: IGI Global. doi:10.4018/978-1-5225-1882-2.ch015

Cipolla-Ficarra, F. V., & Cipolla-Ficarra, M. (2018). Computer Animation for Ingenious Revival. In F. Cipolla-Ficarra, M. Ficarra, M. Cipolla-Ficarra, A. Quiroga, J. Alma, & J. Carré (Eds.), *Technology-Enhanced Human Interaction in Modern Society* (pp. 159–181). Hershey, PA: IGI Global. doi:10.4018/978-1-5225-3437-2.ch008

Cockrell, S., Damron, T. S., Melton, A. M., & Smith, A. D. (2018). Offshoring IT. In M. Khosrow-Pour, D.B.A. (Ed.), Encyclopedia of Information Science and Technology, Fourth Edition (pp. 5476-5489). Hershey, PA: IGI Global. https://doi.org/ doi:10.4018/978-1-5225-2255-3.ch476

Coffey, J. W. (2018). Logic and Proof in Computer Science: Categories and Limits of Proof Techniques. In J. Horne (Ed.), *Philosophical Perceptions on Logic and Order* (pp. 218–240). Hershey, PA: IGI Global. doi:10.4018/978-1-5225-2443-4.ch007

Dale, M. (2017). Re-Thinking the Challenges of Enterprise Architecture Implementation. In M. Tavana (Ed.), *Enterprise Information Systems and the Digitalization of Business Functions* (pp. 205–221). Hershey, PA: IGI Global. doi:10.4018/978-1-5225-2382-6.ch009

Das, A., & Mohanty, M. N. (2021). An Useful Review on Optical Character Recognition for Smart Era Generation. In A. Tyagi (Ed.), *Multimedia and Sensory Input for Augmented, Mixed, and Virtual Reality* (pp. 1–41). IGI Global. https://doi.org/10.4018/978-1-7998-4703-8.ch001

Dash, A. K., & Mohapatra, P. (2021). A Survey on Prematurity Detection of Diabetic Retinopathy Based on Fundus Images Using Deep Learning Techniques. In S. Saxena & S. Paul (Eds.), *Deep Learning Applications in Medical Imaging* (pp. 140–155). IGI Global. https://doi.org/10.4018/978-1-7998-5071-7.ch006

De Maere, K., De Haes, S., & von Kutzschenbach, M. (2017). CIO Perspectives on Organizational Learning within the Context of IT Governance. *International Journal of IT/Business Alignment and Governance, 8*(1), 32-47. https://doi.org/doi:10.4018/IJITBAG.2017010103

Demir, K., Çaka, C., Yaman, N. D., İslamoğlu, H., & Kuzu, A. (2018). Examining the Current Definitions of Computational Thinking. In H. Ozcinar, G. Wong, & H. Ozturk (Eds.), *Teaching Computational Thinking in Primary Education* (pp. 36–64). Hershey, PA: IGI Global. doi:10.4018/978-1-5225-3200-2.ch003

Deng, X., Hung, Y., & Lin, C. D. (2017). Design and Analysis of Computer Experiments. In S. Saha, A. Mandal, A. Narasimhamurthy, S. V, & S. Sangam (Eds.), Handbook of Research on Applied Cybernetics and Systems Science (pp. 264-279). Hershey, PA: IGI Global. doi:10.4018/978-1-5225-2498-4.ch013

Denner, J., Martinez, J., & Thiry, H. (2017). Strategies for Engaging Hispanic/Latino Youth in the US in Computer Science. In Y. Rankin & J. Thomas (Eds.), *Moving Students of Color from Consumers to Producers of Technology* (pp. 24–48). Hershey, PA: IGI Global. doi:10.4018/978-1-5225-2005-4.ch002

Devi, A. (2017). Cyber Crime and Cyber Security: A Quick Glance. In R. Kumar, P. Pattnaik, & P. Pandey (Eds.), *Detecting and Mitigating Robotic Cyber Security Risks* (pp. 160–171). Hershey, PA: IGI Global. doi:10.4018/978-1-5225-2154-9.ch011

Dhaya, R., & Kanthavel, R. (2022). Futuristic Research Perspectives of IoT Platforms. In D. Jeya Mala (Ed.), *Integrating AI in IoT Analytics on the Cloud for Healthcare Applications* (pp. 258–275). IGI Global. doi:10.4018/978-1-7998-9132-1.ch015

Doyle, D. J., & Fahy, P. J. (2018). Interactivity in Distance Education and Computer-Aided Learning, With Medical Education Examples. In M. Khosrow-Pour, D.B.A. (Ed.), Encyclopedia of Information Science and Technology, Fourth Edition (pp. 5829-5840). Hershey, PA: IGI Global. https://doi.org/ doi:10.4018/978-1-5225-2255-3.ch507

Eklund, P. (2021). Reinforcement Learning in Social Media Marketing. In B. Christiansen & T. Škrinjarić (Eds.), *Handbook of Research on Applied AI for International Business and Marketing Applications* (pp. 30–48). IGI Global. https://doi.org/10.4018/978-1-7998-5077-9.ch003

El Ghandour, N., Benaissa, M., & Lebbah, Y. (2021). An Integer Linear Programming-Based Method for the Extraction of Ontology Alignment. *International Journal of Information Technology and Web Engineering, 16*(2), 25–44. https://doi.org/10.4018/IJITWE.2021040102

Elias, N. I., & Walker, T. W. (2017). Factors that Contribute to Continued Use of E-Training among Healthcare Professionals. In F. Topor (Ed.), *Handbook of Research on Individualism and Identity in the Globalized Digital Age* (pp. 403–429). Hershey, PA: IGI Global. doi:10.4018/978-1-5225-0522-8.ch018

Fisher, R. L. (2018). Computer-Assisted Indian Matrimonial Services. In M. Khosrow-Pour, D.B.A. (Ed.), Encyclopedia of Information Science and Technology, Fourth Edition (pp. 4136-4145). Hershey, PA: IGI Global. doi:10.4018/978-1-5225-2255-3.ch358

Galiautdinov, R. (2021). Nonlinear Filtering in Artificial Neural Network Applications in Business and Engineering. In Q. Do (Ed.), *Artificial Neural Network Applications in Business and Engineering* (pp. 1–23). IGI Global. https://doi.org/10.4018/978-1-7998-3238-6.ch001

Gardner-McCune, C., & Jimenez, Y. (2017). Historical App Developers: Integrating CS into K-12 through Cross-Disciplinary Projects. In Y. Rankin & J. Thomas (Eds.), *Moving Students of Color from Consumers to Producers of Technology* (pp. 85–112). Hershey, PA: IGI Global. doi:10.4018/978-1-5225-2005-4.ch005

Garg, P. K. (2021). The Internet of Things-Based Technologies. In S. Kumar, M. Trivedi, P. Ranjan, & A. Punhani (Eds.), *Evolution of Software-Defined Networking Foundations for IoT and 5G Mobile Networks* (pp. 37–65). IGI Global. https://doi.org/10.4018/978-1-7998-4685-7.ch003

Garg, T., & Bharti, M. (2021). Congestion Control Protocols for UWSNs. In N. Goyal, L. Sapra, & J. Sandhu (Eds.), *Energy-Efficient Underwater Wireless Communications and Networking* (pp. 85–100). IGI Global. https://doi.org/10.4018/978-1-7998-3640-7.ch006

Gauttier, S. (2021). A Primer on Q-Method and the Study of Technology. In M. Khosrow-Pour D.B.A. (Eds.), *Encyclopedia of Information Science and Technology, Fifth Edition* (pp. 1746-1756). IGI Global. https://doi.org/10.4018/978-1-7998-3479-3.ch120

Ghafele, R., & Gibert, B. (2018). Open Growth: The Economic Impact of Open Source Software in the USA. In M. Khosrow-Pour (Ed.), *Optimizing Contemporary Application and Processes in Open Source Software* (pp. 164–197). Hershey, PA: IGI Global. doi:10.4018/978-1-5225-5314-4.ch007

Ghobakhloo, M., & Azar, A. (2018). Information Technology Resources, the Organizational Capability of Lean-Agile Manufacturing, and Business Performance. *Information Resources Management Journal, 31*(2), 47–74. doi:10.4018/IRMJ.2018040103

Gikandi, J. W. (2017). Computer-Supported Collaborative Learning and Assessment: A Strategy for Developing Online Learning Communities in Continuing Education. In J. Keengwe & G. Onchwari (Eds.), *Handbook of Research on Learner-Centered Pedagogy in Teacher Education and Professional Development* (pp. 309–333). Hershey, PA: IGI Global. doi:10.4018/978-1-5225-0892-2.ch017

Gokhale, A. A., & Machina, K. F. (2017). Development of a Scale to Measure Attitudes toward Information Technology. In L. Tomei (Ed.), *Exploring the New Era of Technology-Infused Education* (pp. 49–64). Hershey, PA: IGI Global. doi:10.4018/978-1-5225-1709-2.ch004

Goswami, J. K., Jalal, S., Negi, C. S., & Jalal, A. S. (2022). A Texture Features-Based Robust Facial Expression Recognition. *International Journal of Computer Vision and Image Processing, 12*(1), 1–15. https://doi.org/10.4018/IJCVIP.2022010103

Hafeez-Baig, A., Gururajan, R., & Wickramasinghe, N. (2017). Readiness as a Novel Construct of Readiness Acceptance Model (RAM) for the Wireless Handheld Technology. In N. Wickramasinghe (Ed.), *Handbook of Research on Healthcare Administration and Management* (pp. 578–595). Hershey, PA: IGI Global. doi:10.4018/978-1-5225-0920-2.ch035

Hanafizadeh, P., Ghandchi, S., & Asgarimehr, M. (2017). Impact of Information Technology on Lifestyle: A Literature Review and Classification. *International Journal of Virtual Communities and Social Networking*, *9*(2), 1–23. doi:10.4018/IJVCSN.2017040101

Haseski, H. İ., Ilic, U., & Tuğtekin, U. (2018). Computational Thinking in Educational Digital Games: An Assessment Tool Proposal. In H. Ozcinar, G. Wong, & H. Ozturk (Eds.), *Teaching Computational Thinking in Primary Education* (pp. 256–287). Hershey, PA: IGI Global. doi:10.4018/978-1-5225-3200-2.ch013

Hee, W. J., Jalleh, G., Lai, H., & Lin, C. (2017). E-Commerce and IT Projects: Evaluation and Management Issues in Australian and Taiwanese Hospitals. *International Journal of Public Health Management and Ethics*, *2*(1), 69–90. doi:10.4018/IJPHME.2017010104

Hernandez, A. A. (2017). Green Information Technology Usage: Awareness and Practices of Philippine IT Professionals. *International Journal of Enterprise Information Systems*, *13*(4), 90–103. doi:10.4018/IJEIS.2017100106

Hernandez, M. A., Marin, E. C., Garcia-Rodriguez, J., Azorin-Lopez, J., & Cazorla, M. (2017). Automatic Learning Improves Human-Robot Interaction in Productive Environments: A Review. *International Journal of Computer Vision and Image Processing*, *7*(3), 65–75. doi:10.4018/IJCVIP.2017070106

Hirota, A. (2021). Design of Narrative Creation in Innovation: "Signature Story" and Two Types of Pivots. In T. Ogata & J. Ono (Eds.), *Bridging the Gap Between AI, Cognitive Science, and Narratology With Narrative Generation* (pp. 363–376). IGI Global. https://doi.org/10.4018/978-1-7998-4864-6.ch012

Hond, D., Asgari, H., Jeffery, D., & Newman, M. (2021). An Integrated Process for Verifying Deep Learning Classifiers Using Dataset Dissimilarity Measures. *International Journal of Artificial Intelligence and Machine Learning*, *11*(2), 1–21. https://doi.org/10.4018/IJAIML.289536

Horne-Popp, L. M., Tessone, E. B., & Welker, J. (2018). If You Build It, They Will Come: Creating a Library Statistics Dashboard for Decision-Making. In L. Costello & M. Powers (Eds.), *Developing In-House Digital Tools in Library Spaces* (pp. 177–203). Hershey, PA: IGI Global. doi:10.4018/978-1-5225-2676-6.ch009

Hu, H., Hu, P. J., & Al-Gahtani, S. S. (2017). User Acceptance of Computer Technology at Work in Arabian Culture: A Model Comparison Approach. In M. Khosrow-Pour (Ed.), *Handbook of Research on Technology Adoption, Social Policy, and Global Integration* (pp. 205–228). Hershey, PA: IGI Global. doi:10.4018/978-1-5225-2668-1.ch011

Huang, C., Sun, Y., & Fuh, C. (2022). Vehicle License Plate Recognition With Deep Learning. In C. Chen, W. Yang, & L. Chen (Eds.), *Technologies to Advance Automation in Forensic Science and Criminal Investigation* (pp. 161-219). IGI Global. https://doi.org/10.4018/978-1-7998-8386-9.ch009

Ifinedo, P. (2017). Using an Extended Theory of Planned Behavior to Study Nurses' Adoption of Healthcare Information Systems in Nova Scotia. *International Journal of Technology Diffusion*, *8*(1), 1–17. doi:10.4018/IJTD.2017010101

Ilie, V., & Sneha, S. (2018). A Three Country Study for Understanding Physicians' Engagement With Electronic Information Resources Pre and Post System Implementation. *Journal of Global Information Management*, *26*(2), 48–73. doi:10.4018/JGIM.2018040103

Ilo, P. I., Nkiko, C., Ugwu, C. I., Ekere, J. N., Izuagbe, R., & Fagbohun, M. O. (2021). Prospects and Challenges of Web 3.0 Technologies Application in the Provision of Library Services. In M. Khosrow-Pour D.B.A. (Ed.), *Encyclopedia of Information Science and Technology, Fifth Edition* (pp. 1767-1781). IGI Global. https://doi.org/10.4018/978-1-7998-3479-3.ch122

Inoue-Smith, Y. (2017). Perceived Ease in Using Technology Predicts Teacher Candidates' Preferences for Online Resources. *International Journal of Online Pedagogy and Course Design*, *7*(3), 17–28. doi:10.4018/IJOPCD.2017070102

Islam, A. Y. (2017). Technology Satisfaction in an Academic Context: Moderating Effect of Gender. In A. Mesquita (Ed.), *Research Paradigms and Contemporary Perspectives on Human-Technology Interaction* (pp. 187–211). Hershey, PA: IGI Global. doi:10.4018/978-1-5225-1868-6.ch009

Jagdale, S. C., Hable, A. A., & Chabukswar, A. R. (2021). Protocol Development in Clinical Trials for Healthcare Management. In M. Khosrow-Pour D.B.A. (Ed.), *Encyclopedia of Information Science and Technology, Fifth Edition* (pp. 1797-1814). IGI Global. https://doi.org/10.4018/978-1-7998-3479-3.ch124

Jamil, G. L., & Jamil, C. C. (2017). Information and Knowledge Management Perspective Contributions for Fashion Studies: Observing Logistics and Supply Chain Management Processes. In G. Jamil, A. Soares, & C. Pessoa (Eds.), *Handbook of Research on Information Management for Effective Logistics and Supply Chains* (pp. 199–221). Hershey, PA: IGI Global. doi:10.4018/978-1-5225-0973-8.ch011

Jamil, M. I., & Almunawar, M. N. (2021). Importance of Digital Literacy and Hindrance Brought About by Digital Divide. In M. Khosrow-Pour D.B.A. (Ed.), *Encyclopedia of Information Science and Technology, Fifth Edition* (pp. 1683-1698). IGI Global. https://doi.org/10.4018/978-1-7998-3479-3.ch116

Janakova, M. (2018). Big Data and Simulations for the Solution of Controversies in Small Businesses. In M. Khosrow-Pour, D.B.A. (Ed.), Encyclopedia of Information Science and Technology, Fourth Edition (pp. 6907-6915). Hershey, PA: IGI Global. doi:10.4018/978-1-5225-2255-3.ch598

Jhawar, A., & Garg, S. K. (2018). Logistics Improvement by Investment in Information Technology Using System Dynamics. In A. Azar & S. Vaidyanathan (Eds.), *Advances in System Dynamics and Control* (pp. 528–567). Hershey, PA: IGI Global. doi:10.4018/978-1-5225-4077-9.ch017

Kalelioğlu, F., Gülbahar, Y., & Doğan, D. (2018). Teaching How to Think Like a Programmer: Emerging Insights. In H. Ozcinar, G. Wong, & H. Ozturk (Eds.), *Teaching Computational Thinking in Primary Education* (pp. 18–35). Hershey, PA: IGI Global. doi:10.4018/978-1-5225-3200-2.ch002

Kamberi, S. (2017). A Girls-Only Online Virtual World Environment and its Implications for Game-Based Learning. In A. Stricker, C. Calongne, B. Truman, & F. Arenas (Eds.), *Integrating an Awareness of Selfhood and Society into Virtual Learning* (pp. 74–95). Hershey, PA: IGI Global. doi:10.4018/978-1-5225-2182-2.ch006

Kamel, S., & Rizk, N. (2017). ICT Strategy Development: From Design to Implementation – Case of Egypt. In C. Howard & K. Hargiss (Eds.), *Strategic Information Systems and Technologies in Modern Organizations* (pp. 239–257). Hershey, PA: IGI Global. doi:10.4018/978-1-5225-1680-4.ch010

Kamel, S. H. (2018). The Potential Role of the Software Industry in Supporting Economic Development. In M. Khosrow-Pour, D.B.A. (Ed.), Encyclopedia of Information Science and Technology, Fourth Edition (pp. 7259-7269). Hershey, PA: IGI Global. doi:10.4018/978-1-5225-2255-3.ch631

Kang, H., Kang, Y., & Kim, J. (2022). Improved Fall Detection Model on GRU Using PoseNet. *International Journal of Software Innovation, 10*(2), 1–11. https://doi.org/10.4018/IJSI.289600

Kankam, P. K. (2021). Employing Case Study and Survey Designs in Information Research. *Journal of Information Technology Research, 14*(1), 167–177. https://doi.org/10.4018/JITR.2021010110

Karas, V., & Schuller, B. W. (2021). Deep Learning for Sentiment Analysis: An Overview and Perspectives. In F. Pinarbasi & M. Taskiran (Eds.), *Natural Language Processing for Global and Local Business* (pp. 97–132). IGI Global. https://doi.org/10.4018/978-1-7998-4240-8.ch005

Kaufman, L. M. (2022). Reimagining the Magic of the Workshop Model. In T. Driscoll III, (Ed.), *Designing Effective Distance and Blended Learning Environments in K-12* (pp. 89–109). IGI Global. https://doi.org/10.4018/978-1-7998-6829-3.ch007

Kawata, S. (2018). Computer-Assisted Parallel Program Generation. In M. Khosrow-Pour, D.B.A. (Ed.), Encyclopedia of Information Science and Technology, Fourth Edition (pp. 4583-4593). Hershey, PA: IGI Global. doi:10.4018/978-1-5225-2255-3.ch398

Kharb, L., & Singh, P. (2021). Role of Machine Learning in Modern Education and Teaching. In S. Verma & P. Tomar (Ed.), *Impact of AI Technologies on Teaching, Learning, and Research in Higher Education* (pp. 99-123). IGI Global. https://doi.org/10.4018/978-1-7998-4763-2.ch006

Khari, M., Shrivastava, G., Gupta, S., & Gupta, R. (2017). Role of Cyber Security in Today's Scenario. In R. Kumar, P. Pattnaik, & P. Pandey (Eds.), *Detecting and Mitigating Robotic Cyber Security Risks* (pp. 177–191). Hershey, PA: IGI Global. doi:10.4018/978-1-5225-2154-9.ch013

Khekare, G., & Sheikh, S. (2021). Autonomous Navigation Using Deep Reinforcement Learning in ROS. *International Journal of Artificial Intelligence and Machine Learning, 11*(2), 63–70. https://doi.org/10.4018/IJAIML.20210701.oa4

Khouja, M., Rodriguez, I. B., Ben Halima, Y., & Moalla, S. (2018). IT Governance in Higher Education Institutions: A Systematic Literature Review. *International Journal of Human Capital and Information Technology Professionals, 9*(2), 52–67. doi:10.4018/IJHCITP.2018040104

Kiourt, C., Pavlidis, G., Koutsoudis, A., & Kalles, D. (2017). Realistic Simulation of Cultural Heritage. *International Journal of Computational Methods in Heritage Science, 1*(1), 10–40. doi:10.4018/IJCMHS.2017010102

Köse, U. (2017). An Augmented-Reality-Based Intelligent Mobile Application for Open Computer Education. In G. Kurubacak & H. Altinpulluk (Eds.), *Mobile Technologies and Augmented Reality in Open Education* (pp. 154–174). Hershey, PA: IGI Global. doi:10.4018/978-1-5225-2110-5.ch008

Lahmiri, S. (2018). Information Technology Outsourcing Risk Factors and Provider Selection. In M. Gupta, R. Sharman, J. Walp, & P. Mulgund (Eds.), *Information Technology Risk Management and Compliance in Modern Organizations* (pp. 214–228). Hershey, PA: IGI Global. doi:10.4018/978-1-5225-2604-9.ch008

Lakkad, A. K., Bhadaniya, R. D., Shah, V. N., & Lavanya, K. (2021). Complex Events Processing on Live News Events Using Apache Kafka and Clustering Techniques. *International Journal of Intelligent Information Technologies, 17*(1), 39–52. https://doi.org/10.4018/IJIIT.2021010103

Landriscina, F. (2017). Computer-Supported Imagination: The Interplay Between Computer and Mental Simulation in Understanding Scientific Concepts. In I. Levin & D. Tsybulsky (Eds.), *Digital Tools and Solutions for Inquiry-Based STEM Learning* (pp. 33–60). Hershey, PA: IGI Global. doi:10.4018/978-1-5225-2525-7.ch002

Lara López, G. (2021). Virtual Reality in Object Location. In A. Negrón & M. Muñoz (Eds.), *Latin American Women and Research Contributions to the IT Field* (pp. 307–324). IGI Global. https://doi.org/10.4018/978-1-7998-7552-9.ch014

Lee, W. W. (2018). Ethical Computing Continues From Problem to Solution. In M. Khosrow-Pour, D.B.A. (Ed.), Encyclopedia of Information Science and Technology, Fourth Edition (pp. 4884-4897). Hershey, PA: IGI Global. doi:10.4018/978-1-5225-2255-3.ch423

Lin, S., Chen, S., & Chuang, S. (2017). Perceived Innovation and Quick Response Codes in an Online-to-Offline E-Commerce Service Model. *International Journal of E-Adoption, 9*(2), 1–16. doi:10.4018/IJEA.2017070101

Liu, M., Wang, Y., Xu, W., & Liu, L. (2017). Automated Scoring of Chinese Engineering Students' English Essays. *International Journal of Distance Education Technologies, 15*(1), 52–68. doi:10.4018/IJDET.2017010104

Ma, X., Li, X., Zhong, B., Huang, Y., Gu, Y., Wu, M., Liu, Y., & Zhang, M. (2021). A Detector and Evaluation Framework of Abnormal Bidding Behavior Based on Supplier Portrait. *International Journal of Information Technology and Web Engineering, 16*(2), 58–74. https://doi.org/10.4018/IJITWE.2021040104

Mabe, L. K., & Oladele, O. I. (2017). Application of Information Communication Technologies for Agricultural Development through Extension Services: A Review. In T. Tossy (Ed.), *Information Technology Integration for Socio-Economic Development* (pp. 52–101). Hershey, PA: IGI Global. doi:10.4018/978-1-5225-0539-6.ch003

Mahboub, S. A., Sayed Ali Ahmed, E., & Saeed, R. A. (2021). Smart IDS and IPS for Cyber-Physical Systems. In A. Luhach & A. Elçi (Eds.), *Artificial Intelligence Paradigms for Smart Cyber-Physical Systems* (pp. 109–136). IGI Global. https://doi.org/10.4018/978-1-7998-5101-1.ch006

Manogaran, G., Thota, C., & Lopez, D. (2018). Human-Computer Interaction With Big Data Analytics. In D. Lopez & M. Durai (Eds.), *HCI Challenges and Privacy Preservation in Big Data Security* (pp. 1–22). Hershey, PA: IGI Global. doi:10.4018/978-1-5225-2863-0.ch001

Margolis, J., Goode, J., & Flapan, J. (2017). A Critical Crossroads for Computer Science for All: "Identifying Talent" or "Building Talent," and What Difference Does It Make? In Y. Rankin & J. Thomas (Eds.), *Moving Students of Color from Consumers to Producers of Technology* (pp. 1–23). Hershey, PA: IGI Global. doi:10.4018/978-1-5225-2005-4.ch001

Mazzù, M. F., Benetton, A., Baccelloni, A., & Lavini, L. (2022). A Milk Blockchain-Enabled Supply Chain: Evidence From Leading Italian Farms. In P. De Giovanni (Ed.), *Blockchain Technology Applications in Businesses and Organizations* (pp. 73–98). IGI Global. https://doi.org/10.4018/978-1-7998-8014-1.ch004

Mbale, J. (2018). Computer Centres Resource Cloud Elasticity-Scalability (CRECES): Copperbelt University Case Study. In S. Aljawarneh & M. Malhotra (Eds.), *Critical Research on Scalability and Security Issues in Virtual Cloud Environments* (pp. 48–70). Hershey, PA: IGI Global. doi:10.4018/978-1-5225-3029-9.ch003

McKee, J. (2018). The Right Information: The Key to Effective Business Planning. In *Business Architectures for Risk Assessment and Strategic Planning: Emerging Research and Opportunities* (pp. 38–52). Hershey, PA: IGI Global. doi:10.4018/978-1-5225-3392-4.ch003

Meddah, I. H., Remil, N. E., & Meddah, H. N. (2021). Novel Approach for Mining Patterns. *International Journal of Applied Evolutionary Computation, 12*(1), 27–42. https://doi.org/10.4018/IJAEC.2021010103

Mensah, I. K., & Mi, J. (2018). Determinants of Intention to Use Local E-Government Services in Ghana: The Perspective of Local Government Workers. *International Journal of Technology Diffusion, 9*(2), 41–60. doi:10.4018/IJTD.2018040103

Mohamed, J. H. (2018). Scientograph-Based Visualization of Computer Forensics Research Literature. In J. Jeyasekar & P. Saravanan (Eds.), *Innovations in Measuring and Evaluating Scientific Information* (pp. 148–162). Hershey, PA: IGI Global. doi:10.4018/978-1-5225-3457-0.ch010

Montañés-Del Río, M. Á., Cornejo, V. R., Rodríguez, M. R., & Ortiz, J. S. (2021). Gamification of University Subjects: A Case Study for Operations Management. *Journal of Information Technology Research, 14*(2), 1–29. https://doi.org/10.4018/JITR.2021040101

Moore, R. L., & Johnson, N. (2017). Earning a Seat at the Table: How IT Departments Can Partner in Organizational Change and Innovation. *International Journal of Knowledge-Based Organizations*, *7*(2), 1–12. doi:10.4018/IJKBO.2017040101

Mukul, M. K., & Bhattaharyya, S. (2017). Brain-Machine Interface: Human-Computer Interaction. In E. Noughabi, B. Raahemi, A. Albadvi, & B. Far (Eds.), *Handbook of Research on Data Science for Effective Healthcare Practice and Administration* (pp. 417–443). Hershey, PA: IGI Global. doi:10.4018/978-1-5225-2515-8.ch018

Na, L. (2017). Library and Information Science Education and Graduate Programs in Academic Libraries. In L. Ruan, Q. Zhu, & Y. Ye (Eds.), *Academic Library Development and Administration in China* (pp. 218–229). Hershey, PA: IGI Global. doi:10.4018/978-1-5225-0550-1.ch013

Nagpal, G., Bishnoi, G. K., Dhami, H. S., & Vijayvargia, A. (2021). Use of Data Analytics to Increase the Efficiency of Last Mile Logistics for Ecommerce Deliveries. In B. Patil & M. Vohra (Eds.), *Handbook of Research on Engineering, Business, and Healthcare Applications of Data Science and Analytics* (pp. 167–180). IGI Global. https://doi.org/10.4018/978-1-7998-3053-5.ch009

Nair, S. M., Ramesh, V., & Tyagi, A. K. (2021). Issues and Challenges (Privacy, Security, and Trust) in Blockchain-Based Applications. In A. Tyagi, G. Rekha, & N. Sreenath (Eds.), *Opportunities and Challenges for Blockchain Technology in Autonomous Vehicles* (pp. 196–209). IGI Global. https://doi.org/10.4018/978-1-7998-3295-9.ch012

Naomi, J. F. M., K., & V., S. (2021). Machine and Deep Learning Techniques in IoT and Cloud. In S. Velayutham (Ed.), *Challenges and Opportunities for the Convergence of IoT, Big Data, and Cloud Computing* (pp. 225-247). IGI Global. https://doi.org/10.4018/978-1-7998-3111-2.ch013

Nath, R., & Murthy, V. N. (2018). What Accounts for the Differences in Internet Diffusion Rates Around the World? In M. Khosrow-Pour, D.B.A. (Ed.), Encyclopedia of Information Science and Technology, Fourth Edition (pp. 8095-8104). Hershey, PA: IGI Global. https://doi.org/ doi:10.4018/978-1-5225-2255-3.ch705

Nedelko, Z., & Potocan, V. (2018). The Role of Emerging Information Technologies for Supporting Supply Chain Management. In M. Khosrow-Pour, D.B.A. (Ed.), Encyclopedia of Information Science and Technology, Fourth Edition (pp. 5559-5569). Hershey, PA: IGI Global. doi:10.4018/978-1-5225-2255-3.ch483

Negrini, L., Giang, C., & Bonnet, E. (2022). Designing Tools and Activities for Educational Robotics in Online Learning. In N. Eteokleous & E. Nisiforou (Eds.), *Designing, Constructing, and Programming Robots for Learning* (pp. 202–222). IGI Global. https://doi.org/10.4018/978-1-7998-7443-0.ch010

Ngafeeson, M. N. (2018). User Resistance to Health Information Technology. In M. Khosrow-Pour, D.B.A. (Ed.), Encyclopedia of Information Science and Technology, Fourth Edition (pp. 3816-3825). Hershey, PA: IGI Global. doi:10.4018/978-1-5225-2255-3.ch331

Nguyen, T. T., Giang, N. L., Tran, D. T., Nguyen, T. T., Nguyen, H. Q., Pham, A. V., & Vu, T. D. (2021). A Novel Filter-Wrapper Algorithm on Intuitionistic Fuzzy Set for Attribute Reduction From Decision Tables. *International Journal of Data Warehousing and Mining, 17*(4), 67–100. https://doi.org/10.4018/IJDWM.2021100104

Nigam, A., & Dewani, P. P. (2022). Consumer Engagement Through Conditional Promotions: An Exploratory Study. *Journal of Global Information Management, 30*(5), 1–19. https://doi.org/10.4018/JGIM.290364

Odagiri, K. (2017). Introduction of Individual Technology to Constitute the Current Internet. In *Strategic Policy-Based Network Management in Contemporary Organizations* (pp. 20–96). Hershey, PA: IGI Global. doi:10.4018/978-1-68318-003-6.ch003

Odia, J. O., & Akpata, O. T. (2021). Role of Data Science and Data Analytics in Forensic Accounting and Fraud Detection. In B. Patil & M. Vohra (Eds.), *Handbook of Research on Engineering, Business, and Healthcare Applications of Data Science and Analytics* (pp. 203–227). IGI Global. https://doi.org/10.4018/978-1-7998-3053-5.ch011

Okike, E. U. (2018). Computer Science and Prison Education. In I. Biao (Ed.), *Strategic Learning Ideologies in Prison Education Programs* (pp. 246–264). Hershey, PA: IGI Global. doi:10.4018/978-1-5225-2909-5.ch012

Olelewe, C. J., & Nwafor, I. P. (2017). Level of Computer Appreciation Skills Acquired for Sustainable Development by Secondary School Students in Nsukka LGA of Enugu State, Nigeria. In C. Ayo & V. Mbarika (Eds.), *Sustainable ICT Adoption and Integration for Socio-Economic Development* (pp. 214–233). Hershey, PA: IGI Global. doi:10.4018/978-1-5225-2565-3.ch010

Oliveira, M., Maçada, A. C., Curado, C., & Nodari, F. (2017). Infrastructure Profiles and Knowledge Sharing. *International Journal of Technology and Human Interaction, 13*(3), 1–12. doi:10.4018/IJTHI.2017070101

Otarkhani, A., Shokouhyar, S., & Pour, S. S. (2017). Analyzing the Impact of Governance of Enterprise IT on Hospital Performance: Tehran's (Iran) Hospitals – A Case Study. *International Journal of Healthcare Information Systems and Informatics*, *12*(3), 1–20. doi:10.4018/IJHISI.2017070101

Otunla, A. O., & Amuda, C. O. (2018). Nigerian Undergraduate Students' Computer Competencies and Use of Information Technology Tools and Resources for Study Skills and Habits' Enhancement. In M. Khosrow-Pour, D.B.A. (Ed.), Encyclopedia of Information Science and Technology, Fourth Edition (pp. 2303-2313). Hershey, PA: IGI Global. https://doi.org/ doi:10.4018/978-1-5225-2255-3.ch200

Özçınar, H. (2018). A Brief Discussion on Incentives and Barriers to Computational Thinking Education. In H. Ozcinar, G. Wong, & H. Ozturk (Eds.), *Teaching Computational Thinking in Primary Education* (pp. 1–17). Hershey, PA: IGI Global. doi:10.4018/978-1-5225-3200-2.ch001

Pandey, J. M., Garg, S., Mishra, P., & Mishra, B. P. (2017). Computer Based Psychological Interventions: Subject to the Efficacy of Psychological Services. *International Journal of Computers in Clinical Practice*, *2*(1), 25–33. doi:10.4018/IJCCP.2017010102

Pandkar, S. D., & Paatil, S. D. (2021). Big Data and Knowledge Resource Centre. In S. Dhamdhere (Ed.), *Big Data Applications for Improving Library Services* (pp. 90–106). IGI Global. https://doi.org/10.4018/978-1-7998-3049-8.ch007

Patro, C. (2017). Impulsion of Information Technology on Human Resource Practices. In P. Ordóñez de Pablos (Ed.), *Managerial Strategies and Solutions for Business Success in Asia* (pp. 231–254). Hershey, PA: IGI Global. doi:10.4018/978-1-5225-1886-0.ch013

Patro, C. S., & Raghunath, K. M. (2017). Information Technology Paraphernalia for Supply Chain Management Decisions. In M. Tavana (Ed.), *Enterprise Information Systems and the Digitalization of Business Functions* (pp. 294–320). Hershey, PA: IGI Global. doi:10.4018/978-1-5225-2382-6.ch014

Paul, P. K. (2018). The Context of IST for Solid Information Retrieval and Infrastructure Building: Study of Developing Country. *International Journal of Information Retrieval Research*, *8*(1), 86–100. doi:10.4018/IJIRR.2018010106

Paul, P. K., & Chatterjee, D. (2018). iSchools Promoting "Information Science and Technology" (IST) Domain Towards Community, Business, and Society With Contemporary Worldwide Trend and Emerging Potentialities in India. In M. Khosrow-Pour, D.B.A. (Ed.), Encyclopedia of Information Science and Technology, Fourth Edition (pp. 4723-4735). Hershey, PA: IGI Global. https://doi.org/ doi:10.4018/978-1-5225-2255-3.ch410

Pessoa, C. R., & Marques, M. E. (2017). Information Technology and Communication Management in Supply Chain Management. In G. Jamil, A. Soares, & C. Pessoa (Eds.), *Handbook of Research on Information Management for Effective Logistics and Supply Chains* (pp. 23–33). Hershey, PA: IGI Global. doi:10.4018/978-1-5225-0973-8.ch002

Pineda, R. G. (2018). Remediating Interaction: Towards a Philosophy of Human-Computer Relationship. In M. Khosrow-Pour (Ed.), *Enhancing Art, Culture, and Design With Technological Integration* (pp. 75–98). Hershey, PA: IGI Global. doi:10.4018/978-1-5225-5023-5.ch004

Prabha, V. D., & R., R. (2021). Clinical Decision Support Systems: Decision-Making System for Clinical Data. In G. Rani & P. Tiwari (Eds.), *Handbook of Research on Disease Prediction Through Data Analytics and Machine Learning* (pp. 268-280). IGI Global. https://doi.org/10.4018/978-1-7998-2742-9.ch014

Pushpa, R., & Siddappa, M. (2021). An Optimal Way of VM Placement Strategy in Cloud Computing Platform Using ABCS Algorithm. *International Journal of Ambient Computing and Intelligence*, *12*(3), 16–38. https://doi.org/10.4018/IJACI.2021070102

Qian, Y. (2017). Computer Simulation in Higher Education: Affordances, Opportunities, and Outcomes. In P. Vu, S. Fredrickson, & C. Moore (Eds.), *Handbook of Research on Innovative Pedagogies and Technologies for Online Learning in Higher Education* (pp. 236–262). Hershey, PA: IGI Global. doi:10.4018/978-1-5225-1851-8.ch011

Rahman, N. (2017). Lessons from a Successful Data Warehousing Project Management. *International Journal of Information Technology Project Management*, *8*(4), 30–45. doi:10.4018/IJITPM.2017100103

Rahman, N. (2018). Environmental Sustainability in the Computer Industry for Competitive Advantage. In M. Khosrow-Pour (Ed.), *Green Computing Strategies for Competitive Advantage and Business Sustainability* (pp. 110–130). Hershey, PA: IGI Global. doi:10.4018/978-1-5225-5017-4.ch006

Rajh, A., & Pavetic, T. (2017). Computer Generated Description as the Required Digital Competence in Archival Profession. *International Journal of Digital Literacy and Digital Competence*, *8*(1), 36–49. doi:10.4018/IJDLDC.2017010103

Raman, A., & Goyal, D. P. (2017). Extending IMPLEMENT Framework for Enterprise Information Systems Implementation to Information System Innovation. In M. Tavana (Ed.), *Enterprise Information Systems and the Digitalization of Business Functions* (pp. 137–177). Hershey, PA: IGI Global. doi:10.4018/978-1-5225-2382-6.ch007

Rao, A. P., & Reddy, K. S. (2021). Automated Soil Residue Levels Detecting Device With IoT Interface. In V. Sathiyamoorthi & A. Elci (Eds.), *Challenges and Applications of Data Analytics in Social Perspectives* (Vol. S, pp. 123–135). IGI Global. https://doi.org/10.4018/978-1-7998-2566-1.ch007

Rao, Y. S., Rauta, A. K., Saini, H., & Panda, T. C. (2017). Mathematical Model for Cyber Attack in Computer Network. *International Journal of Business Data Communications and Networking*, *13*(1), 58–65. doi:10.4018/IJBDCN.2017010105

Rapaport, W. J. (2018). Syntactic Semantics and the Proper Treatment of Computationalism. In M. Danesi (Ed.), *Empirical Research on Semiotics and Visual Rhetoric* (pp. 128–176). Hershey, PA: IGI Global. doi:10.4018/978-1-5225-5622-0.ch007

Raut, R., Priyadarshinee, P., & Jha, M. (2017). Understanding the Mediation Effect of Cloud Computing Adoption in Indian Organization: Integrating TAM-TOE- Risk Model. *International Journal of Service Science, Management, Engineering, and Technology*, *8*(3), 40–59. doi:10.4018/IJSSMET.2017070103

Rezaie, S., Mirabedini, S. J., & Abtahi, A. (2018). Designing a Model for Implementation of Business Intelligence in the Banking Industry. *International Journal of Enterprise Information Systems*, *14*(1), 77–103. doi:10.4018/IJEIS.2018010105

Rezende, D. A. (2018). Strategic Digital City Projects: Innovative Information and Public Services Offered by Chicago (USA) and Curitiba (Brazil). In M. Lytras, L. Daniela, & A. Visvizi (Eds.), *Enhancing Knowledge Discovery and Innovation in the Digital Era* (pp. 204–223). Hershey, PA: IGI Global. doi:10.4018/978-1-5225-4191-2.ch012

Rodriguez, A., Rico-Diaz, A. J., Rabuñal, J. R., & Gestal, M. (2017). Fish Tracking with Computer Vision Techniques: An Application to Vertical Slot Fishways. In M. S., & V. V. (Eds.), Multi-Core Computer Vision and Image Processing for Intelligent Applications (pp. 74-104). Hershey, PA: IGI Global. https://doi.org/doi:10.4018/978-1-5225-0889-2.ch003

Romero, J. A. (2018). Sustainable Advantages of Business Value of Information Technology. In M. Khosrow-Pour, D.B.A. (Ed.), Encyclopedia of Information Science and Technology, Fourth Edition (pp. 923-929). Hershey, PA: IGI Global. doi:10.4018/978-1-5225-2255-3.ch079

Romero, J. A. (2018). The Always-On Business Model and Competitive Advantage. In N. Bajgoric (Ed.), *Always-On Enterprise Information Systems for Modern Organizations* (pp. 23–40). Hershey, PA: IGI Global. doi:10.4018/978-1-5225-3704-5.ch002

Rosen, Y. (2018). Computer Agent Technologies in Collaborative Learning and Assessment. In M. Khosrow-Pour, D.B.A. (Ed.), Encyclopedia of Information Science and Technology, Fourth Edition (pp. 2402-2410). Hershey, PA: IGI Global. doi:10.4018/978-1-5225-2255-3.ch209

Roy, D. (2018). Success Factors of Adoption of Mobile Applications in Rural India: Effect of Service Characteristics on Conceptual Model. In M. Khosrow-Pour (Ed.), *Green Computing Strategies for Competitive Advantage and Business Sustainability* (pp. 211–238). Hershey, PA: IGI Global. doi:10.4018/978-1-5225-5017-4.ch010

Ruffin, T. R., & Hawkins, D. P. (2018). Trends in Health Care Information Technology and Informatics. In M. Khosrow-Pour, D.B.A. (Ed.), Encyclopedia of Information Science and Technology, Fourth Edition (pp. 3805-3815). Hershey, PA: IGI Global. doi:10.4018/978-1-5225-2255-3.ch330

Sadasivam, U. M., & Ganesan, N. (2021). Detecting Fake News Using Deep Learning and NLP. In S. Misra, C. Arumugam, S. Jaganathan, & S. S. (Eds.), *Confluence of AI, Machine, and Deep Learning in Cyber Forensics* (pp. 117-133). IGI Global. https://doi.org/10.4018/978-1-7998-4900-1.ch007

Safari, M. R., & Jiang, Q. (2018). The Theory and Practice of IT Governance Maturity and Strategies Alignment: Evidence From Banking Industry. *Journal of Global Information Management, 26*(2), 127–146. doi:10.4018/JGIM.2018040106

Sahin, H. B., & Anagun, S. S. (2018). Educational Computer Games in Math Teaching: A Learning Culture. In E. Toprak & E. Kumtepe (Eds.), *Supporting Multiculturalism in Open and Distance Learning Spaces* (pp. 249–280). Hershey, PA: IGI Global. doi:10.4018/978-1-5225-3076-3.ch013

Sakalle, A., Tomar, P., Bhardwaj, H., & Sharma, U. (2021). Impact and Latest Trends of Intelligent Learning With Artificial Intelligence. In S. Verma & P. Tomar (Eds.), *Impact of AI Technologies on Teaching, Learning, and Research in Higher Education* (pp. 172-189). IGI Global. https://doi.org/10.4018/978-1-7998-4763-2.ch011

Sala, N. (2021). Virtual Reality, Augmented Reality, and Mixed Reality in Education: A Brief Overview. In D. Choi, A. Dailey-Hebert, & J. Estes (Eds.), *Current and Prospective Applications of Virtual Reality in Higher Education* (pp. 48–73). IGI Global. https://doi.org/10.4018/978-1-7998-4960-5.ch003

Salunkhe, S., Kanagachidambaresan, G., Rajkumar, C., & Jayanthi, K. (2022). Online Detection and Prediction of Fused Deposition Modelled Parts Using Artificial Intelligence. In S. Salunkhe, H. Hussein, & J. Davim (Eds.), *Applications of Artificial Intelligence in Additive Manufacturing* (pp. 194–209). IGI Global. https://doi.org/10.4018/978-1-7998-8516-0.ch009

Samy, V. S., Pramanick, K., Thenkanidiyoor, V., & Victor, J. (2021). Data Analysis and Visualization in Python for Polar Meteorological Data. *International Journal of Data Analytics*, *2*(1), 32–60. https://doi.org/10.4018/IJDA.2021010102

Sanna, A., & Valpreda, F. (2017). An Assessment of the Impact of a Collaborative Didactic Approach and Students' Background in Teaching Computer Animation. *International Journal of Information and Communication Technology Education*, *13*(4), 1–16. doi:10.4018/IJICTE.2017100101

Sarivougioukas, J., & Vagelatos, A. (2022). Fused Contextual Data With Threading Technology to Accelerate Processing in Home UbiHealth. *International Journal of Software Science and Computational Intelligence*, *14*(1), 1–14. https://doi.org/10.4018/IJSSCI.285590

Scott, A., Martin, A., & McAlear, F. (2017). Enhancing Participation in Computer Science among Girls of Color: An Examination of a Preparatory AP Computer Science Intervention. In Y. Rankin & J. Thomas (Eds.), *Moving Students of Color from Consumers to Producers of Technology* (pp. 62–84). Hershey, PA: IGI Global. doi:10.4018/978-1-5225-2005-4.ch004

Shanmugam, M., Ibrahim, N., Gorment, N. Z., Sugu, R., Dandarawi, T. N., & Ahmad, N. A. (2022). Towards an Integrated Omni-Channel Strategy Framework for Improved Customer Interaction. In P. Lai (Ed.), *Handbook of Research on Social Impacts of E-Payment and Blockchain Technology* (pp. 409–427). IGI Global. https://doi.org/10.4018/978-1-7998-9035-5.ch022

Sharma, A., & Kumar, S. (2021). Network Slicing and the Role of 5G in IoT Applications. In S. Kumar, M. Trivedi, P. Ranjan, & A. Punhani (Eds.), *Evolution of Software-Defined Networking Foundations for IoT and 5G Mobile Networks* (pp. 172–190). IGI Global. https://doi.org/10.4018/978-1-7998-4685-7.ch010

Siddoo, V., & Wongsai, N. (2017). Factors Influencing the Adoption of ISO/IEC 29110 in Thai Government Projects: A Case Study. *International Journal of Information Technologies and Systems Approach, 10*(1), 22–44. doi:10.4018/IJITSA.2017010102

Silveira, C., Hir, M. E., & Chaves, H. K. (2022). An Approach to Information Management as a Subsidy of Global Health Actions: A Case Study of Big Data in Health for Dengue, Zika, and Chikungunya. In J. Lima de Magalhães, Z. Hartz, G. Jamil, H. Silveira, & L. Jamil (Eds.), *Handbook of Research on Essential Information Approaches to Aiding Global Health in the One Health Context* (pp. 219–234). IGI Global. https://doi.org/10.4018/978-1-7998-8011-0.ch012

Simões, A. (2017). Using Game Frameworks to Teach Computer Programming. In R. Alexandre Peixoto de Queirós & M. Pinto (Eds.), *Gamification-Based E-Learning Strategies for Computer Programming Education* (pp. 221–236). Hershey, PA: IGI Global. doi:10.4018/978-1-5225-1034-5.ch010

Simões de Almeida, R., & da Silva, T. (2022). AI Chatbots in Mental Health: Are We There Yet? In A. Marques & R. Queirós (Eds.), *Digital Therapies in Psychosocial Rehabilitation and Mental Health* (pp. 226–243). IGI Global. https://doi.org/10.4018/978-1-7998-8634-1.ch011

Singh, L. K., Khanna, M., Thawkar, S., & Gopal, J. (2021). Robustness for Authentication of the Human Using Face, Ear, and Gait Multimodal Biometric System. *International Journal of Information System Modeling and Design, 12*(1), 39–72. https://doi.org/10.4018/IJISMD.2021010103

Sllame, A. M. (2017). Integrating LAB Work With Classes in Computer Network Courses. In H. Alphin Jr, R. Chan, & J. Lavine (Eds.), *The Future of Accessibility in International Higher Education* (pp. 253–275). Hershey, PA: IGI Global. doi:10.4018/978-1-5225-2560-8.ch015

Smirnov, A., Ponomarev, A., Shilov, N., Kashevnik, A., & Teslya, N. (2018). Ontology-Based Human-Computer Cloud for Decision Support: Architecture and Applications in Tourism. *International Journal of Embedded and Real-Time Communication Systems, 9*(1), 1–19. doi:10.4018/IJERTCS.2018010101

Smith-Ditizio, A. A., & Smith, A. D. (2018). Computer Fraud Challenges and Its Legal Implications. In M. Khosrow-Pour, D.B.A. (Ed.), Encyclopedia of Information Science and Technology, Fourth Edition (pp. 4837-4848). Hershey, PA: IGI Global. doi:10.4018/978-1-5225-2255-3.ch419

Sosnin, P. (2018). Figuratively Semantic Support of Human-Computer Interactions. In *Experience-Based Human-Computer Interactions: Emerging Research and Opportunities* (pp. 244–272). Hershey, PA: IGI Global. doi:10.4018/978-1-5225-2987-3.ch008

Srilakshmi, R., & Jaya Bhaskar, M. (2021). An Adaptable Secure Scheme in Mobile Ad hoc Network to Protect the Communication Channel From Malicious Behaviours. *International Journal of Information Technology and Web Engineering*, *16*(3), 54–73. https://doi.org/10.4018/IJITWE.2021070104

Sukhwani, N., Kagita, V. R., Kumar, V., & Panda, S. K. (2021). Efficient Computation of Top-K Skyline Objects in Data Set With Uncertain Preferences. *International Journal of Data Warehousing and Mining*, *17*(3), 68–80. https://doi.org/10.4018/IJDWM.2021070104

Susanto, H., Yie, L. F., Setiana, D., Asih, Y., Yoganingrum, A., Riyanto, S., & Saputra, F. A. (2021). Digital Ecosystem Security Issues for Organizations and Governments: Digital Ethics and Privacy. In Z. Mahmood (Ed.), *Web 2.0 and Cloud Technologies for Implementing Connected Government* (pp. 204–228). IGI Global. https://doi.org/10.4018/978-1-7998-4570-6.ch010

Syväjärvi, A., Leinonen, J., Kivivirta, V., & Kesti, M. (2017). The Latitude of Information Management in Local Government: Views of Local Government Managers. *International Journal of Electronic Government Research*, *13*(1), 69–85. doi:10.4018/IJEGR.2017010105

Tanque, M., & Foxwell, H. J. (2018). Big Data and Cloud Computing: A Review of Supply Chain Capabilities and Challenges. In A. Prasad (Ed.), *Exploring the Convergence of Big Data and the Internet of Things* (pp. 1–28). Hershey, PA: IGI Global. doi:10.4018/978-1-5225-2947-7.ch001

Teixeira, A., Gomes, A., & Orvalho, J. G. (2017). Auditory Feedback in a Computer Game for Blind People. In T. Issa, P. Kommers, T. Issa, P. Isaías, & T. Issa (Eds.), *Smart Technology Applications in Business Environments* (pp. 134–158). Hershey, PA: IGI Global. doi:10.4018/978-1-5225-2492-2.ch007

Tewari, P., Tiwari, P., & Goel, R. (2022). Information Technology in Supply Chain Management. In V. Garg & R. Goel (Eds.), *Handbook of Research on Innovative Management Using AI in Industry 5.0* (pp. 165–178). IGI Global. https://doi.org/10.4018/978-1-7998-8497-2.ch011

Thompson, N., McGill, T., & Murray, D. (2018). Affect-Sensitive Computer Systems. In M. Khosrow-Pour, D.B.A. (Ed.), Encyclopedia of Information Science and Technology, Fourth Edition (pp. 4124-4135). Hershey, PA: IGI Global. doi:10.4018/978-1-5225-2255-3.ch357

Triberti, S., Brivio, E., & Galimberti, C. (2018). On Social Presence: Theories, Methodologies, and Guidelines for the Innovative Contexts of Computer-Mediated Learning. In M. Marmon (Ed.), *Enhancing Social Presence in Online Learning Environments* (pp. 20–41). Hershey, PA: IGI Global. doi:10.4018/978-1-5225-3229-3.ch002

Tripathy, B. K. T. R., S., & Mohanty, R. K. (2018). Memetic Algorithms and Their Applications in Computer Science. In S. Dash, B. Tripathy, & A. Rahman (Eds.), Handbook of Research on Modeling, Analysis, and Application of Nature-Inspired Metaheuristic Algorithms (pp. 73-93). Hershey, PA: IGI Global. https://doi.org/ doi:10.4018/978-1-5225-2857-9.ch004

Turulja, L., & Bajgoric, N. (2017). Human Resource Management IT and Global Economy Perspective: Global Human Resource Information Systems. In M. Khosrow-Pour (Ed.), *Handbook of Research on Technology Adoption, Social Policy, and Global Integration* (pp. 377–394). Hershey, PA: IGI Global. doi:10.4018/978-1-5225-2668-1.ch018

Unwin, D. W., Sanzogni, L., & Sandhu, K. (2017). Developing and Measuring the Business Case for Health Information Technology. In K. Moahi, K. Bwalya, & P. Sebina (Eds.), *Health Information Systems and the Advancement of Medical Practice in Developing Countries* (pp. 262–290). Hershey, PA: IGI Global. doi:10.4018/978-1-5225-2262-1.ch015

Usharani, B. (2022). House Plant Leaf Disease Detection and Classification Using Machine Learning. In M. Mundada, S. Seema, S. K.G., & M. Shilpa (Eds.), *Deep Learning Applications for Cyber-Physical Systems* (pp. 17-26). IGI Global. https://doi.org/10.4018/978-1-7998-8161-2.ch002

Vadhanam, B. R. S., M., Sugumaran, V., V., V., & Ramalingam, V. V. (2017). Computer Vision Based Classification on Commercial Videos. In M. S., & V. V. (Eds.), Multi-Core Computer Vision and Image Processing for Intelligent Applications (pp. 105-135). Hershey, PA: IGI Global. https://doi.org/ doi:10.4018/978-1-5225-0889-2.ch004

Vairinho, S. (2022). Innovation Dynamics Through the Encouragement of Knowledge Spin-Off From Touristic Destinations. In C. Ramos, S. Quinteiro, & A. Gonçalves (Eds.), *ICT as Innovator Between Tourism and Culture* (pp. 170–190). IGI Global. https://doi.org/10.4018/978-1-7998-8165-0.ch011

Valverde, R., Torres, B., & Motaghi, H. (2018). A Quantum NeuroIS Data Analytics Architecture for the Usability Evaluation of Learning Management Systems. In S. Bhattacharyya (Ed.), *Quantum-Inspired Intelligent Systems for Multimedia Data Analysis* (pp. 277–299). Hershey, PA: IGI Global. doi:10.4018/978-1-5225-5219-2.ch009

Vassilis, E. (2018). Learning and Teaching Methodology: "1:1 Educational Computing. In K. Koutsopoulos, K. Doukas, & Y. Kotsanis (Eds.), *Handbook of Research on Educational Design and Cloud Computing in Modern Classroom Settings* (pp. 122–155). Hershey, PA: IGI Global. doi:10.4018/978-1-5225-3053-4.ch007

Verma, S., & Jain, A. K. (2022). A Survey on Sentiment Analysis Techniques for Twitter. In B. Gupta, D. Peraković, A. Abd El-Latif, & D. Gupta (Eds.), *Data Mining Approaches for Big Data and Sentiment Analysis in Social Media* (pp. 57–90). IGI Global. https://doi.org/10.4018/978-1-7998-8413-2.ch003

Wang, H., Huang, P., & Chen, X. (2021). Research and Application of a Multidimensional Association Rules Mining Method Based on OLAP. *International Journal of Information Technology and Web Engineering, 16*(1), 75–94. https://doi.org/10.4018/IJITWE.2021010104

Wexler, B. E. (2017). Computer-Presented and Physical Brain-Training Exercises for School Children: Improving Executive Functions and Learning. In B. Dubbels (Ed.), *Transforming Gaming and Computer Simulation Technologies across Industries* (pp. 206–224). Hershey, PA: IGI Global. doi:10.4018/978-1-5225-1817-4.ch012

Wimble, M., Singh, H., & Phillips, B. (2018). Understanding Cross-Level Interactions of Firm-Level Information Technology and Industry Environment: A Multilevel Model of Business Value. *Information Resources Management Journal, 31*(1), 1–20. doi:10.4018/IRMJ.2018010101

Wimmer, H., Powell, L., Kilgus, L., & Force, C. (2017). Improving Course Assessment via Web-based Homework. *International Journal of Online Pedagogy and Course Design, 7*(2), 1–19. doi:10.4018/IJOPCD.2017040101

Wong, S. (2021). Gendering Information and Communication Technologies in Climate Change. In M. Khosrow-Pour D.B.A. (Eds.), *Encyclopedia of Information Science and Technology, Fifth Edition* (pp. 1408-1422). IGI Global. https://doi.org/10.4018/978-1-7998-3479-3.ch096

Wong, Y. L., & Siu, K. W. (2018). Assessing Computer-Aided Design Skills. In M. Khosrow-Pour, D.B.A. (Ed.), Encyclopedia of Information Science and Technology, Fourth Edition (pp. 7382-7391). Hershey, PA: IGI Global. doi:10.4018/978-1-5225-2255-3.ch642

Wongsurawat, W., & Shrestha, V. (2018). Information Technology, Globalization, and Local Conditions: Implications for Entrepreneurs in Southeast Asia. In P. Ordóñez de Pablos (Ed.), *Management Strategies and Technology Fluidity in the Asian Business Sector* (pp. 163–176). Hershey, PA: IGI Global. doi:10.4018/978-1-5225-4056-4.ch010

Yamada, H. (2021). Homogenization of Japanese Industrial Technology From the Perspective of R&D Expenses. *International Journal of Systems and Service-Oriented Engineering*, *11*(2), 24–51. doi:10.4018/IJSSOE.2021070102

Yang, Y., Zhu, X., Jin, C., & Li, J. J. (2018). Reforming Classroom Education Through a QQ Group: A Pilot Experiment at a Primary School in Shanghai. In H. Spires (Ed.), *Digital Transformation and Innovation in Chinese Education* (pp. 211–231). Hershey, PA: IGI Global. doi:10.4018/978-1-5225-2924-8.ch012

Yilmaz, R., Sezgin, A., Kurnaz, S., & Arslan, Y. Z. (2018). Object-Oriented Programming in Computer Science. In M. Khosrow-Pour, D.B.A. (Ed.), Encyclopedia of Information Science and Technology, Fourth Edition (pp. 7470-7480). Hershey, PA: IGI Global. doi:10.4018/978-1-5225-2255-3.ch650

Yu, L. (2018). From Teaching Software Engineering Locally and Globally to Devising an Internationalized Computer Science Curriculum. In S. Dikli, B. Etheridge, & R. Rawls (Eds.), *Curriculum Internationalization and the Future of Education* (pp. 293–320). Hershey, PA: IGI Global. doi:10.4018/978-1-5225-2791-6.ch016

Yuhua, F. (2018). Computer Information Library Clusters. In M. Khosrow-Pour, D.B.A. (Ed.), Encyclopedia of Information Science and Technology, Fourth Edition (pp. 4399-4403). Hershey, PA: IGI Global. doi:10.4018/978-1-5225-2255-3.ch382

Zakaria, R. B., Zainuddin, M. N., & Mohamad, A. H. (2022). Distilling Blockchain: Complexity, Barriers, and Opportunities. In P. Lai (Ed.), *Handbook of Research on Social Impacts of E-Payment and Blockchain Technology* (pp. 89–114). IGI Global. https://doi.org/10.4018/978-1-7998-9035-5.ch007

Zhang, Z., Ma, J., & Cui, X. (2021). Genetic Algorithm With Three-Dimensional Population Dominance Strategy for University Course Timetabling Problem. *International Journal of Grid and High Performance Computing*, *13*(2), 56–69. https://doi.org/10.4018/IJGHPC.2021040104

About the Contributors

U. Vignesh is currently an Assistant Professor Senior Grade 2 in School of Computer Science and Engineering, Vellore Institute of Technology (VIT) - Chennai. Prior to his recent appointment at the VIT, he was a Post-Doctoral Fellow in National Institute of Technology (NIT), Trichy – India. Dr. Vignesh received his undergraduate degree in B.Tech (IT) as well as his M.Tech (IT) degree from Anna University - Chennai, and his PhD in Computer Science and Engineering from VIT University - Chennai. Dr. Vignesh published several papers in preferred Journals, patents and chapters in books, and participated in a range of forums on computer science, social science, etc. He also presented various academic as well as research-based papers at several national and international conferences. His research activities are currently twofold: while the first research activity is set to explore the developmental role that society needs with technology such as, Artificial Intelligence; the second major research theme that he is pursuing is focused on the bioinformatics and data mining.

Rahul Ratnakumar received his Bachelor's degree from College of Engineering Trivandrum, University of Kerala in Electrical and Electronics Engineering in 2009. He obtained Master of Engineering degree from Birla Institute of Technology and Science (BITS), Pilani with a specialization in Microelectronics in 2012. He worked on the project- Interface circuit design for a MEMS Vibratory Gyroscope, with the Analog and Mixed Signal Design group of CSIR-CEERI Pilani, India. He is completed his doctoral work in Malaviya National Institute of Technology (MNIT), Jaipur, India in the topic of VLSI implementation of Clustering Algorithms. He has a post-doctoral fellowship from the Department of Systems and Computer Engineering at Carleton University, Canada in the field of Autonomous Vehicles from Jan to June 2021. He is currently working as an Assistant Professor in the Department of Electronics and Communication Engg., Manipal Institute of Technology (MIT), Manipal Academy of Higher Education (MAHE), Manipal, Karnataka, India. His field of Interest are Digital VLSI architecture, Bio-inspired systems, Image Processing, System on Chip Design, Soft computing, Machine and Deep Learning.

* * *

Aritro Dey received his BE, ME, and PhD in Electrical Engineering from Jadavpur University, Kolkata, India, in 2008, 2011 and 2017 respectively. He has worked as Scientist / Engineer 'C' at National Control Law team, Aeronautical Development Agency, Ministry of Defence, Bangalore, India from 2015 to 2018. He is working as an Assistant Professor at the Department of Electrical Engineering, National Institute of Technology Durgapur, India from December, 2018. His research interests include nonlinear state estimation, statistical signal processing, robust control and designing control law for aerospace applications.

Ezhilarasan K. holds a UG degree from Ayya Nadar Janaki Ammal College, Sivakasi, Tamil Nadu, India, which he earned in 2005. He obtained a Master of Computer Applications (M.C.A) degree from M. Kumarasamy Engineering College, Karur, Tamil Nadu, India in 2008. In 2016, he completed his Ph.D. at Gandhigram Rural Institute (Deemed University), Dindigul, Tamil Nadu, India. Since 2016, he has been serving as a Guest Teacher at the Centre for Geoinformatics, Gandhigram Rural Institute (Deemed University), Dindigul, Tamil Nadu, India. Additionally, he gained experience as a Project Fellow under the UGC-Major Research Project from July 2010 to January 2013. He was also the recipient of a fellowship from the Jawaharlal Nehru Memorial Fund for Doctoral Studies (JNMF) from January 2014 to June 2015. His primary research interests revolve around the fields of Medical Image Processing and Satellite Image Processing. K. Ezhilarasan has shared his research findings through publications in national and international conferences and journals. Furthermore, he holds the esteemed position of a lifetime member of the Indian Society For Technical Education (ISTE).

Muthukumar M. is an Assistant Professor with an academic background comprising M.Sc., M.Tech., and Ph.D. degrees. He brings a wealth of experience to the field, with a professional journey spanning over 12 years. His expertise lies in the realm of "Applications of Geospatial Technologies in Natural Resources & Natural Disaster Mapping, Management & Mitigation." Dr. Muthukumar has actively contributed to scholarly publications, including one chapter in a book and a notable number of papers in edited volumes and proceedings. His research output extends to both international and national journals, showcasing his dedication to advancing knowledge in the field. Throughout his career, Dr. Muthukumar has been involved in major projects, demonstrating his commitment to practical applications. Notably, he has successfully handled projects with a significant financial outlay, totaling Rs. 80.34 Lakhs during the period 2010-2015. In addition to his research endeavors, Dr. Muthukumar has actively participated in academic events. He has organized three seminars, conferences, or workshops and attended a noteworthy 17 such events. His influence extends to the international stage, with ten invited talks at national and international conferences. Dr. Muthukumar's global engagement includes a visit to Norway, contributing to the exchange of ideas and collaborations. Beyond his research and academic contributions, Dr. Muthukumar has taken on additional responsibilities, reflecting his leadership qualities. These include serving as Residential Warden (2010-2013), Deputy Warden (2013 – till date), and NSS Programme Officer (since 13.03.15). Dr. M. Muthukumar's multifaceted contributions underscore his commitment to advancing geospatial technologies and his impact on academia and disaster management.

Chitra P. is a highly dedicated and accomplished Assistant Professor in Computer Science with a passion for fostering student learning and contributing to cutting-edge research. Proven track record of excellence in teaching, curriculum development, and mentorship. Specialized expertise in [specific areas of computer science, e.g., artificial intelligence, machine learning, cybersecurity]. Committed to creating an inclusive and innovative learning environment. Actively engaged in research, with numerous publications in reputable journals and conferences.

Biswapratim Roy received his B.Tech from University of Engineering & Management, Kolkata, and M.Tech from Netaji Subhash Engineering College(under MAKAUT), in 2020 and 2022 respectively. Currently he is pursuing his PhD from National Institute of Technology Durgapur. His research interests include linear and nonlinear control, fuzzy logic based control, robust control and designing control law for aerospace applications.

Karthigai Selvi S. is serving as an Assistant Professor in the School of Computer Science and Engineering at Galgotias University, Uttar Pradesh. She had been serving as a Guest Teacher in the Department of Computer Science and Applications, The Gandhigram Rural Institute – Deemed to be University, India for last five years. She did her UG (B.Sc. Mathematics) and Ph.D (Fulltime) in The Gandhigram Rural Institute – Deemed to be University. She has more than Ten years Teaching experiences, One year corporate experience like Programmer and served as Project Consultant in Mother Teresa Women's University, India. Her research work focuses on Human Brain Image Processing and Sustainable Development. She bagged a best presenter award in 4th World Environment Summit on 2023 and a National Award in Student Research Convention held on 2013. She has published more than thirty papers in reputed journals (SCI, Scopus and Web of Science), International and National Conferences. She has served as a Resource Person in Universities and Colleges.

Index

H

M

N

P

R

S

www.ingramcontent.com/pod-product-compliance
Lightning Source LLC
LaVergne TN
LVHW080648011026
849160LV00014B/1362
* 9 7 9 8 3 6 9 3 4 4 6 2 0 *